MOLECULAR BIOLOGY
OF HUMAN
HEPATITIS VIRUSES

MOLECULAR BIOLOGY OF HUMAN HEPATITIS VIRUSES

J. Monjardino

Imperial College School of Medicine
St. Mary's Campus

Imperial College Press

Published by

Imperial College Press
203 Electrical Engineering Building
Imperial College
London SW7 2BT

Distributed by

World Scientific Publishing Co. Pte. Ltd.
P O Box 128, Farrer Road, Singapore 912805
USA office: Suite 1B, 1060 Main Street, River Edge, NJ 07661
UK office: 57 Shelton Street, Covent Garden, London WC2H 9HE

British Library Cataloguing-in-Publication Data
A catalogue record for this book is available from the British Library.

MOLECULAR BIOLOGY OF HUMAN HEPATITIS VIRUSES

ISBN   1-86094-048-X

Printed in Singapore by Uto-Print

To Maria Emilia for her support and fortitude during
the preparation of this book.

Preface

Why Hepatitis Viruses?

It will be obvious to the reader that no essentially virological criterion could have brought together a collection of viruses so disparate that it includes a picornavirus (hepatitis A virus), a hepadnavirus (hepatitis B virus), a pestivirus or flavivirus-related virus (hepatitis C virus), a viroid/satellite RNA-like virus (hepatitis Delta virus), and a calici- or related virus (hepatitis E virus). Instead what these diverse viruses have in common is their ability to cause hepatitis in man, similar if not indistinguishable in both its clinical manifestations and histological presentation.

Hepatitis viruses have been the object of intense study in the last twenty years and a wealth of information has accumulated relating to their molecular structures and biological life cycles. Besides the medical implications of such findings, some of the research carried out on these fascinating agents has also generated much new information that transcends their immediate relevance to the understanding of the particular virus and which has proved to be of major general biological significance.

For a variety of reasons, i.e., narrow host range, inefficient or non-existent cell culture systems for virus propagation and low virus titres both in tissues and in biological fluids, conventional virological experimental approaches have so far proved to be ineffective in the study of these viruses. In their place, researchers have had to make full use of their ingenuity and of the entire modern technological armamentarium offered by genetic engineering and molecular biology. This they have been able to do with singular success as we hope this text will illustrate.

Some clinical information has also been included. Although not meant to be comprehensive, since this is not a clinical text, it is in our view a valuable addition which contributes to a better overall understanding in areas which are closer or of more immediate relevance to the molecular studies.

I am greatly indebted to Drs. B. Bottcher and A. Crowther, MRC Cambridge, Prof. Dr. G. Siegl, IKMI, St. Gallen, Switzerland, Dr. K. Krawczynski and Dr. Charles D. Humphrey from CDC, Atlanta, USA and Professor M. Balayan, Institute for Poliomyelitis and Virological Studies, Moscow, Russia for providing illustrations. Finally I also wish to thank Mrs A. Wadmore (ICSM/ St. Mary's) for doing the artwork.

J.M. March 1997

Contents

PART 1

HEPATITIS A VIRUS

History

The recorded history of epidemic jaundice goes back to the 5th century BC, with epidemics associated with wars and famine during the 7th and 8th centuries AD, and with major outbreaks in Europe during the 17th and 18th centuries (J. Melnick, 1995) . During this period it was often confused with leptospirosis. During the 19th century, acute catarrhal jaundice was often recorded as a mild disease thought to be due to the obstruction of the common bile duct (Melnick, 1995). Its infectious nature only became recognized in 1923 when it was first demonstrated that infectious jaundice was the epidemic form of acute catarrhal jaundice.

Although the first unmistakeable outbreaks of serum hepatitis were recorded in 1885 among workers of the Bremen shipyard and the inmates of a Merzig asylum for the mentally insane who had been vaccinated against smallpox, the infectious nature of such a disease and the distinction between the two major types, infectious and serum hepatitis, or A and B, as they have since been designated, was only incontrovertibly demonstrated much later with the studies of Krugman at the Willowbrook State School for mentally handicapped children (Krugman, 1962; 1967). Such investigations, some of which of questionable ethical design, led to the identification of two types of hepatitis, each caused by a different agent and showing no heterologous immunity. This finding explained the observed occurrence of a second infection in most inmates and the duplication of the infection in subjects described as volunteers; it also documented both the passive immunity against these agents and the active immune protection against hepatitis B virus, later to lead to the development of the vaccine against this agent. The demonstration by Deinhardt in 1967 that marmoset monkeys are susceptible to hepatitis A agent further confirmed its identity as a separate virus.

Well before the Hepatitis A Virus (HAV) was first isolated, transmitted to laboratory animals or characterized by electron microscopy (EM) much information had already been gathered about the disease and its agent. Precise details describing the signs and symptoms of hepatitis were already published in disease reports of the late fifties. Also known were the incubation period (15–40 days), the acute onset, its preferential seasonal incidence during autumn/winter and its highest incidence amongst children and young adults. Information on the agent related to its size, since it could pass through a Seitz EK bacterial filter, to its presence in feces during the incubation and acute phases and in blood 3 days before onset and during acute phase, to the route of infection which was mostly oral, and to the effective protection provided by gamma globulin.

HAV was first seen by EM in 1973 (Feinstone *et al.*) in the feces of volunteers and subsequently classified as a separate genus of the Picornaviridae. Serological assays for detection of HAV antigen and antibody were first described in 1975 by Provost *et al.* The virus was first reported to grow in cultured primate cell lines by Provost *et al.* in 1979, an observation which eventually led to the development of a an effective vaccine in 1992.

The Virus

Classification: The Picornaviridae

HAV is a member of the Picornaviridae family. These viruses are particles of icosahedral symmetry and about 30 nanometers diameter with a genome of single-stranded plus polarity RNA of M_r of approximately 2.5×10^6. Replication occurs in the cytoplasm of the infected cell.

Within this family, HAV constitutes a separate genus, *Hepatovirus*, which includes four additional genera, i.e., *Enterovirus* (poliovirus), *Rhinovirus* (human rhinoviruses), *Aphtovirus* (foot-and-mouth disease virus) and *Cardiovirus* (encephalomyocarditis virus of mice (EMC) as well as some viruses of insects).

The Virion

The virion is a naked spherical particle of 27 nm diameter (Fig. 1) which bands between 1.32 and 1.34 g/ml in CsCl and sediments at 156–160S. Other infectious particles of higher (1.42 g/ml, 'heavy') and lower (1.24 g/ml, 'light') densities are also present in small numbers and empty particles (density: 1.29–1.31 gm/ml) can account for 60% of the total number of particles present. Defective interfering (DI) particles are also found which contain defective genomes deleted in both the structural and non-structural regions and capable of interfering with virus replication.

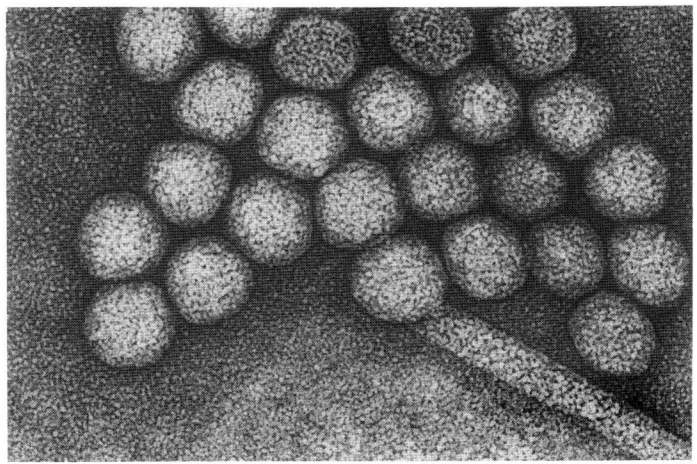

Fig. 1 Electron micrograph of HAV purified from human faeces. Magnification: × 390,000. (Courtesy of Prof. G. Siegl).

The mature infectious particle assembles 12 pentameric units and each of these dissociates into five protomers making a total of 60 protomers with a triangulation number (T) of 1. Each protomer, in turn, is made up of one molecule each of the structural proteins VP1, VP2 and VP3 with VP4 probably buried inside the particle and not part of the shell. When compared to other picornaviruses it is unusually resistant to environmental factors being stable within a wide pH range (3–10), sustaining a temperature of 60° for 30 minutes at pH:7.0 with no changes in sedimentation behaviour, antigenicity or EM morphological appearance (Siegl *et al.*, 1984a) and showing high resistance to chlorination (Peterson *et al.*, 1983).

HAV Genome

The RNA genome is single-stranded, of plus polarity and about 7.500 nucleotides long, with a buoyant density of 1.64 g/ml and a sedimentation of 33S. From several full sequences which have been published since 1983 (Ticehurst *et al.*), it is known that the genome comprises a 5′ terminal sequence of 733 nucleotides which is not translated, a long Open Reading Frame (ORF) of 6,681 nucleotides and a 3′ untranslated terminal region of 64 nucleotides. In addition there is a poly adenylic acid tail 40–80 nucleotides long at the 3′ end.

The 5′ terminus is uncapped but covalently linked to a small peptide VPg (Virus protein genome-linked).

The overall GC content of 38% is significantly lower than in other members of the Picornavirus family with which there is low overall nucleotide homology. The sequence shares with that of rhinoviruses a preference for A and U in the third codon position.

Comparative analysis of sequences derived from many different geographical locations has shown little genetic variation. All human and simian HAV isolates are now classified into seven distinct genotypes which differ by 15%–25% of their nucleotide sequence. Subtypes within these groups differ by approximately 7.5% in nucleotide sequence and individual subtypes or those occurring in epidemics or endemic areas show less than 3% diversity over periods of up to 15 years. During a particular infection, no nucleotide changes were detected over a 41 day period and only one change was detected in culture over 72 consecutive passages (Seigl 1993). This genetic stability contrasts with the genetic changes reported in other picornaviruses where 1–2 nucleotide changes a week have been reported in poliovirus.

The positive sense polarity of the HAV genome was established after transfection of virion-derived RNA into cultured cells as well as by its *in vitro* translation (Locarnini *et al.*, 1981; Gauss-Muller *et al.*, 1984; Cohen *et al.*, 1987a).

HAV Proteins

The single ORF can code for a polyprotein of 2,227 aminoacids and it corresponds to a total of 11 proteins in three different regions, P1–P3 (Fig. 2). As with nucleotides, amino acid homology with other members of the family is low (up to 29%), predominantly associated with proteins involved in replication and polyprotein processing.

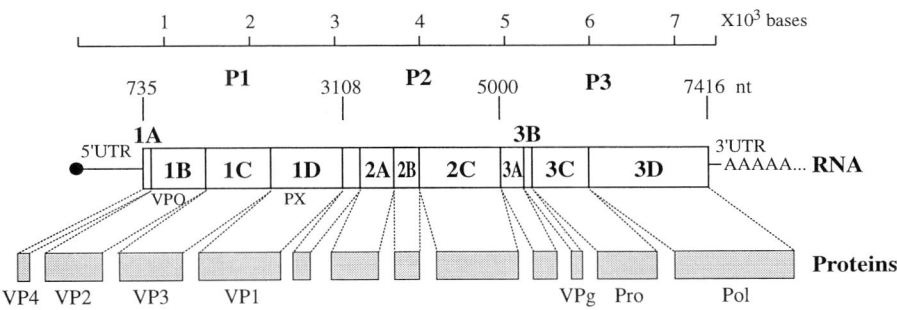

Fig. 2 HAV genomic and translation maps. 5′ UTR: 5′ Untranslated Region; 3′ UTR: 3′ Untranslated Region. VP1, 2, 3 and 4: Capsid proteins.VPg: Genome-linked viral protein; Pro: Protease; Pol: Polymerase.

Region P1 comprises the structural proteins VP1–VP4 although VP4, in spite of being reported in several studies , has not been conclusively identified. As to the three structural proteins of the capsid, VPO, VP3 and VP1, five copies appear to interact to form a precursor particle or pentamer, and 12 pentamers, or 60 copies of the main structural proteins, assemble into empty capsid particles which become infectious after the RNA genome is packaged.

Region P2 encodes three non-structural proteins, 2A–C. The function of 2A appears to be different in different members of the Picornavirus family. It appears to be a protease in entero and rhinoviruses but an amino acid sequence carrying no protease consensus sequences in either cardiovirus or aphtovirus members where it is a 16 kD protein of unknown function in the former and a remnant of less than 20 amino acids, also of unknown function, in the latter. HAV 2A shows little similarity to the 2A regions of members of other picornaviruses and even its precise boundaries (and therefore size) have not yet been precisely established. After primary polyprotein cleavage 2A remains covalently linked to capsid protein VP1 (see below). Recently it has been shown that deletions of 30–45 nucleotides (10–15 aas) from the presumed central portion of 2A appear to have little or no effect on viral growth, both in cultured cells and in marmoset liver(Harmon *et al.*,1995)

Region P3 encodes four additional non-structural proteins, 3A–D. The 3C product is a cysteine protease which also shares properties with serine proteases of the Chymotrypsin family (Allaire *et al.*,1994). In common with other picornavirus proteases; the cleavage it mediates occurs primarily after a glutamine residue; but its specificity regarding the P1′ position appears to be less stringent in that it can include Met or Val residues, in addition to Gly and Ser which are found in other members of the family. The catalytic site probably involves only the diad (and not triad as in chymotrypsin-like proteases) of Cys 172, which is the nucleophile, and His 44 which is a general base, and the 3C specificity for glutamine is determined by His 191 (Allaire *et al.*, 1994).

Similarly to what has been shown in the papain family of enzymes, substitution of the P1 Glu residue for the amino aldehyde in the preferred HAV 3C protease peptide substrates has resulted in a potent enzyme inhibitor capable of discriminating between picornavirus proteinases with only slightly different optimum substrates (Malcolm *et al.*, 1995).

The 3D protein has been predicted, by analogy with several other picornaviruses, to correspond to the enzyme RNA polymerase. The protein was shown to be cleaved autocatalytically by 3C protease between glutamine and arginine (aas 1738 and 17390) when efficiently expressed in BS-C-1 cells, but remained insoluble and enzymatically inactive. This in contrast with the 3D protein of poliovirus which remains partly soluble and shows marked polymerase activity in a similar system (Tesar *et al.*, 1994).

Molecular Biology of HAV Replication

Cell Attachment and Virus Entry

Although cell culture systems are now available for the study of HAV infection little is known about the initial stages of infection, i.e., adsorption of the virus to the cell, virus internalization and uncoating, or the molecular mechanism of genome replication. A very recent report describes the identification of a glycoprotein on the surface of African Green Monkey cells which is recognized by several infection-blocking monoclonal antibodies but which provides only partial susceptibility to HAV once expressed in HAV non-susceptible mouse LTκ⁻ cells (Kaplan *et al.*, 1996). Whether such a protein is a functional receptor, an attachment receptor or a protein sited close to the functional receptor (with the monoclonals blocking infection by steric hindrance) remains to be established. Replication occurs in the cytoplasm of the infected cell and little is known about the assembly of new virions and their release.

Regulatory Sequences in 5′ and 3′ UTRs

All picornaviruses have long and highly structured 5′ UTRs which constitute approximately 10% of the genome. These 5' UTRs are well conserved within each group but different between groups. Computer-predicted folding of these 5' UTRs indicate that rhinovirus and enterovirus form similar structures (type 1), and cardiovirus and aphtovirus form a different structure (type 2). It is presently unclear whether hepatoviruses contain a structure which is related to type 2 or a new structure altogether.

The complex folding of the 5′ UTR displays regulatory determinants involved in essential viral functions (Harmon *et al.*, 1991). Such domains, particularly those located near the 5′ terminus are thought to be recognized by specific factors, host and viral, that are involved in RNA replication. In addition, sequences that direct cap-independent translation, the internal ribosome entry site (IRES),have also been described downstream from the 5′ terminus and extending over several hundred nucleotides. Functional independence of the 5′ terminal structures ('cloverleaf') involved in replication from the IRES involved in translation has been demonstrated in several reports although recently questioned.

In the case of HAV the 5'UTR has not been characterized as well as in other picornaviruses. The internal ribosome entry site (IRES) or ribosome landing pad has been located between nucleotides 161 and 734 of the 5′ UTR and facilitates the binding of the ribosome onto an uncapped messenger RNA with initiation of translation several hundred of nucleotides downstream from the 5′ terminus (Jackson, 1988). Experiments where chimaeric RNA was generated by replacing HAV IRES with EMCV IRES produced viable virus in similar yields to HAV prototype (although improved translation *in vitro*) and showed that more than 150 of the most 5′ terminal nucleotides are required for viral replication (Jia *et al.*, 1996). Mutational and amino acid sequence analysis have shown that translation is initiated at the 11th or 12th initiation codon (AUG) at positions 735–737 or 741–743. Each codon by itself appears to be sufficient for synthesis and the 12th codon is preferentially used in BSC-1 infected cells. Little information is available about the significance of the remaining 5′ UTR, although the terminal 5′ UUC has been shown to be essential for infection and a single base change at position 687 (U → G) was reported to be associated with increased virus yield by infected BSC-1 cells.

By contrast, with the 5′ UTR where the sequence is highly conserved amongst isolates (95%), the 3′ UTR shows a much higher degree of variability. Little is known about the function of 3′ UTR but currently proposed mechanisms of virus replication would predict that regulatory sequences involved in the control of replication would be contained within this region. Detailed analysis has defined a short, possibly *cis*-acting, element which is 23 nucleotides long and which appears to bind proteins during the establishment and progression of infection *in vitro* and a significantly homologous (79%) element has been reported in poliovirus. Subsequent studies have indicated that this effect may be associated with viral persistence *in vitro* of non-cytolytic strains of HAV.

Kinetics of Viral Replication

When compared with other picornaviruses, HAV replicates poorly in cultured cells with a protracted initial lag period , extended replicative phase and low yield of virions (Locarnini *et al.*, 1981; Siegl *et al.*, 1984b). Host cell metabolism is not inhibited in contrast to what is found for other picornaviruses.

Most virus strains produce a non-cytolytic infection of cultured cells. One-step replication analysis of non-cytolytic infection is initiated by a lag-phase of 1–2 days followed in sequence by the initiation of minus RNA synthesis, plus-strand synthesis, synthesis of viral proteins and ultimately the production of virions which is most active for a period of 3–6 days. A second phase follows which is characterized by down regulation of RNA synthesis but maintained synthesis of viral structural proteins and finally a third phase of persistent infection defined by minimal viral replication. By contrast cytolytic strains induce a short lag-phase of 6–12 hours, with maximal viral RNA replication occurring at 24 hour p.i. and exponentially production of progeny virus up to day 4 p.i. with cell-lysis detected between 3–9 days after infection.

The study of molecular factors determining virus replication was carried out by sequence comparisons between strains barely replicating and either the ones emerging from these through serial passage that were active replicators or those determining persistence/cytolysis. Results are still not conclusive although several base changes have been identified in 5′ UTR, 2B and 2C (Cohen *et al.*, 1987b, c; Jansen *et al.*, 1988; Cohen *et al.*, 1989). However none of these mutations individually was able to produce the change in viral growth and interaction with other mutations (which were variable) was required. Host cells were also shown to be determinant in HAV replication and factors involved were found not to be cell-cycle related.

The underlying molecular mechanism of the striking switch off of virus replication leading on to viral persistence remains unclear although the reduced synthesis of minus strand appears to immediately precede the metabolic shutdown.

Polyprotein Processing

The polyprotein encoded in the long ORF undergoes precisely ordered processing during the life cycle of the virus. Investigation of the proteolytic events in picornaviruses has shown that it first occurs co-translationally at the P1/P2 boundary and is then carried out in *cis* by the 2A protease in entero and rhinoviruses or at the 2A/2B junction by an intrinsic junctional autocatalytic activity in cardio- and aphtoviruses. Next the polyprotein is further processed by the protease 3C , or proteolytic precursor 3CD, either in *cis* or *trans*. Finally P0 is processed to generate P2 and P4 during the maturation of the particle.

By contrast, in HAV a single protease encoded in 3C appears to be responsible for all the proteolytic events involved in the processing of the viral polyprotein (Fig. 3). Recent studies based on the *in vitro* processing of labelled polypeptides and of full length *in vitro* synthesized polyprotein have shown that purified recombinant 3C protease is capable of recognizing the

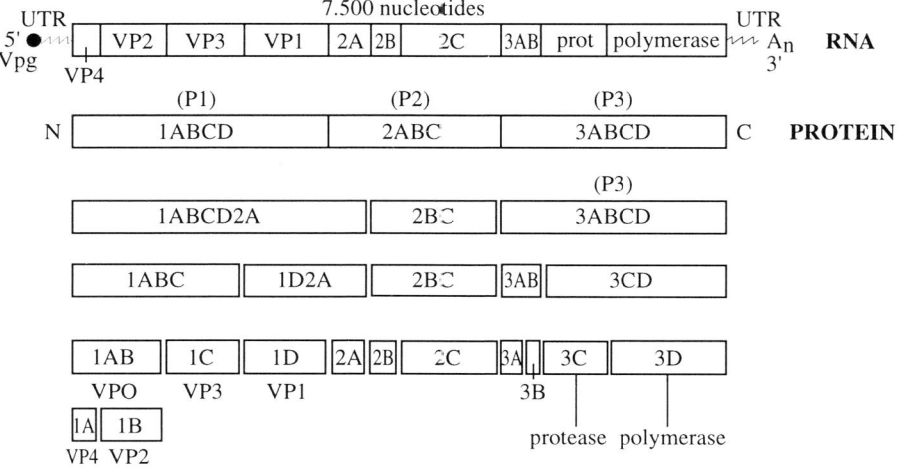

Fig. 3   Sequential processing of HAV polyprotein. 5' and 3' UTRs: Untranslated regions. A_n: Poly A tail; VP1, 2, 3, and 4: Capsid proteins; VP0: Capsid precursor of VP1 and VP4.

nine putative cleavage sites and have established a hierarchy of preferred sites (Schultheiss *et al.*, 1994; Schultheiss *et al.*, 1995). Of these the first to be cleaved are those at the P1–2A/2B (as indicated in Fig. 3) and 2C/3A boundaries, in good agreement with translation kinetics studies, where intramolecular proteolysis by a *cis*-operating protease generated P3 as the first detectable intermediate (Jia *et al.*, 1991; Harmon *et al.*, 1992; Jurgensen *et al*, 1993; Schultheiss *et al.*, 1994). Reports have shown that after primary polyprotein cleavage 2A remains covalently linked to capsid protein VP1 and that the PV1–2A intermediate (PX) accumulates as a major product until late in morphogenesis when it is finally cleaved (Fig. 3) (Anderson and Ross 1990; Borvec and Anderson 1993). Thus, the first products appear to be P1–2A, P3 and 2BC with cleavage of 2BC and processing of P1–2A occurring later. Processing by 3C proteinase within P1 and P3 was less efficient. These products were produced by intermolecular cleavage of an *in vitro* translated polypeptide precursor VP1–P2–P3 (made defective in its protease) by a recombinant 3C proteinase in *trans* and have been shown to be identical to those generated by intramolecular proteolysis of the wild type polypeptide and comparable to products processed from the complete HAV polyprotein (Schultheiss *et al.*, 1994). The products generated from processing within the structural region, VP0, VP3 and VP1–2A also appeared to correspond to the equivalent polypetides found in HAV-infected cells.

Antigenic Properties of HAV

The antigenic organization of HAV is uncertain in spite of a strong neutralizing antibody response having been well documented for a long time. One immunodominant neutralizing site which is discontinuous and involves VP3 and VP1 has been well characterized (Ping *et al.*, 1988). Three distinct VP1–VP3 epitopes have been identified in the pentamers of the HAV structural unit with two additional epitopes being generated as the result of the assembly of the capsid. T cell response, including the mapping of epitopes, has been insufficiently characterized.

The Disease

Hepatitis A or infectious hepatitis is a self-limiting infection of the liver caused by HAV. The virus is usually water-borne and trasmitted by the faeco-oral route as a result of ingestion of viral particles present in faeces through contaminated food, water or hands.

Infected food is commonly traced to an infectious food-handler, to contaminated water used to wash uncooked foods or to shellfish infected by sewage-contaminated sea water. Parenteral transmission via infectious blood and blood products is very rare as is sexual transmission.

Natural History of Hepatitis A

After an incubation period of between 15–50 days, average one month, there are prodromal symptoms of malaise, fatigue and loss of apetite, and fever in some cases. This is followed by jaundice and dark urine about two weeks after the onset of clinical disease. During this period both the liver enzymes and bile pigments are raised in serum. However in some cases, common amongst children but rare in adults, the disease is asymptomatic.

The patient is infectious between two and five weeks after exposure while the virus, which is secreted in the bile, is present in the stools.

The infection is self-limiting with ultimate clearance of the virus. This occurs before the jaundice disappears and is associated with a strong neutralizing antibody response which confers life-long protection. The antibody response includes IgM, IgA and IgG. IgM can be detected days before the onset of symptoms and disappears by three months whereas IgG, being the primary defense against re-infection, becomes detectable within three months and persists throughout life.

The strictly hepatotropic virus is not directly cytopathic and liver damage appears to occur as the result of the cellular immune response directed at the infected liver cells. Although the histological changes observed in the infected liver are common to other forms of viral hepatitis some features are suggestive of HAV infection. These include the periportal distribution of the inflammatory changes with bile stasis (cholestasis) and prominent hepatocyte damage with multivesicular bodies in the cells showing cytopathic changes.

The virus gets across the intestinal wall and reaches the liver by a mechanism which has not been fully elucidated. Replication in the liver is accompanied by virus release into the blood stream and excretion into the bile and faeces.

Epidemiology

The ocurrence of hepatitis A is closely related to socio-economic development with high incidence, endemic areas corresponding to the poorest regions of the world. With an estimated number of new cases of the order of 1,400,000 a year worldwide the incidence per continental region varies between 10 cases/100,000/year in the United States and 20–40 or 20–60 in either Central and South America or Africa and Middle East respectively.

In the poorest regions where the infection is endemic, contact with the virus occurs early in life and by the age of five more than 90% of children have been infected.

As socio-economic conditions improve infection occurs first in increasingly older age groups but with similar overall prevalence; but as conditions improve still further (Japan, Northern Europe), infection is almost exclusively seen in adults and overall prevalence begins to decline to such an extent that the majority of the population has no markers of infection. Contact with the virus amongst these populations results normally from visits to regions of high endemicity and the illness tends to have a more serious course.

Prevention

Passive immunization following administration of HAV immunoglobulins is very effective, providing 85% protection and lasting for three months. It is effective when administered prior to exposure or even within two weeks after exposure to the virus as shown in field trials carried out more than five decades ago (Havens and Paul 1945). This observation led to the development of a vaccine which was first reported by Provost and Hileman in 1978 when they achieved successful immunoprotection of marmosets against challenge with live HAV following immunization with formaldehyde-inactivated virus prepared from infected liver.

A vaccine is presently commercially available which consists of formaldehyde-killed virus. It is based on the attenuated and cell-culture adapted HM175 HAV strain which is grown in human diploid MRC-5 cells. The manufacturing process is laborious in view of the low virus yield in cell culture and this is reflected in the unit cost. Two or three injections are required to provide adequate long-lasting protection.

In addition, two attenuated live vaccines have been licensed in China and one based on an inactivated virosome (reconstituted influenza virosome as carrier) has been developed and marketed in Switzerland.

General consensus appears to favour selective administration of the vaccine in view of the self-limiting, very rarely fulminant course of the acute infection with HAV. However, HAV still shows a relatively high prevalence even in developed countries like the USA, where 50% of the cases of viral hepatitis are caused by HAV and where antibody prevalence is still as high as 38.2% in the general population, reflecting marked social-economic inequalities. In view of this and of the more severe course of the infection in adults a case has also been advanced for wider, if not universal, administration of the vaccine.

Attempts at producing a live-attenuated vaccine, which would hopefully require one injection, or better still one oral dose, for adequate protection and thus reduce administration costs whilst increasing long-term efficacy have so far proved unsuccessful in Western countries (although two have been licensed in China). The virus required as the basis for such a vaccine must be grown efficiently in cell culture, be of attenuated virulence, and strongly immunogenic. Whereas genetic determinants for *in vitro* culture both in MRC-5 and AGMK cells have been precisely mapped in the 5' UTR (Funkhouse *et al.*, 1996), simultaneous down-regulation of virulence whilst retaining strong immunogenicity remains a major challenge.

Some Unanswered Questions

Understanding the biological cycle of HAV at the molecular level would be greatly advanced by the availability of an efficient cell culture system for virus propagation. Such a system would also contribute significantly to vaccine production and generate detailed information on virion polypeptide composition and architecture.

Further genome mutation analysis will be required for clarification at the molecular level of HAV adaptation to *in vitro* cell culture growth, and *in vivo* viral replication and pathogenicity

In spite of significant progress achieved in relation to the processing of the HAV polyprotein little is known about HAV RNA replication. Both the development of a more efficient cell culture system and the ability to study replication *in vitro* with recombinant RNA polymerase and synthetic genomic RNA should produce important information in this area.

The neutralizing immune response against HAV is also still not completely understood. Neutralizing antibodies need to be incontrovertibly defined in terms of the antigens involved and the major epitope(s) precisely mapped. The role of cell immunity in HAV neutralization and its involvement in pathogenesis remains to be elucidated.

Finally in the area of prevention the ultimate objective must be the complete eradication of the virus, a predictably feasible goal since the virus appears not to have an animal reservoir. In order to attain such a goal, a new attenuated live vaccine for single oral administration and providing lasting protection would prove invaluable.

References

Allaire, M., Chernaia, M.M., Malcolm, B.A. and James, M.N. *Nature* 1994; **369**: 72–76.

Anderson, D.A. and Ross, B.C. *J. Virol.* 1990; **64**: 5284–5289.

Borvec, S.V. and Anderson, D.A. *J. Virol.* 1993; **67**: 3095–3102.

Cohen, J.R., Rosenblum, B., Ticehurst, J.R. *et al.* *PNAS* 1987a; **84**: 2497–2501.

Cohen, J.I., Ticehurst, J.R., Feinstone, S.M. *et al.* *J. Virol.* 1987b; **61**: 3035–3039.

Cohen, J.R., Ticehurst, J.R., Purcell, R.H. *et al.* *J. Virol.* 1987c; **61**: 50–59.

Cohen, J.R., Rosenblum, B., Feinstone, S.M. *et al.* *J. Virol.* 1989; **63**: 5364–5370.

Deinhard, F., Holmes, A.W., Capps, R.B. and Popper, H.J. *Exp. Med.* 1967; **125**: 673–688.

Feinstone, S.M., Kapikian, A.Z. and Purcell, R.H. *Science* 1973; **182**:1026–1028.

Funkhouse, A.W., Raychaudhuri, G., Purcell, R.H. *et al.* *J. Virol.* 1996; **70**: 7948–7957

Gauss-Muller, V., von-der-Helm, K. and Deinhardt, F. *Virology* 1984; **137**: 182–184.

Harmon, S.A., Richards, O.C., Summers, D.F. and Ehrenfeld, E. *J. Virol.*1991; **65**: 2757–2760.

Harmon, S.A., Updike, Jia X.Y. *et al.* *J.Virol.* 1992; **66**: 5242–5247.

Harmon, S.A., Emerson, S.U., Huang, Y.K. *et al.* *J. Virol.* 1995; **69**: 5576–5581.

Havens, W.P. Jr and Paul, J.R. *JAMA* 1945; **129**: 270–272.

Jackson, R.J. *Nature* 1988; **334**: 292–294.

Jansen, R.W., Newbold, J.E. and Lemon, S. *Virology* 1988; **163**: 299–307.

Jia, X-Y., Ehrenfeld, E. and Summers, D.F. *J. Virol.* 1991; **65**: 2595–2600.

Jia, X-Y., Tesar, M., Summers, D. and Ehrenfeld, E. *J. Virol.* 1996; **70**: 2861–2868.

Jurgensen, D., Kusov, Y.Y., Facke, M. *et al.* *J. Gen. Virol.* 1993; **74**: 677–683.

Kaplan, G., Totsuka, A., Thompson, P. *et al.* *EMBO J.* 1996; **15**: 4282–4296.

Krugman, S., Ward, R. and Giles, J.P. *Am. J. Med.* 1962; **32**: 717–728.

Krugman, S., Giles, J.P. and Hammond, J. *JAMA* 1967; **200**: 365–373.

Locarnini, S.A., Coulepis, A.G., Westway, E.G. and Gust, I.D. *J. Virol.* 1981; **37**: 216–225.

Malcolm, B.A., Lowe, C., Shechosky, S. *et al.* *Biochemistry* 1995; **34**: 8172–8179.

Melnick, J.L. *J. Infect. Dis.* 1995; **171** Suppl. 1: S2–8.

Peterson, D.A., Hurley, T.R., Hoff, J.C. and Wolfe, L.G. *Appl. Environ. Microbiol.* 1983; **45**: 223–227.

Ping, L-H., Jansen, R.W., Stapleton, J.T. *et al.* PNAS 1988; **85**: 8281–8285.

Provost, P.J., Wolamski, B.S. and Miller, W.J. *et al.* *Amer. J. Med. Sci.* 1975; **270**: 87–92.

Provost, P.J. and Hilleman, M.R. *Proc. Soc. Exp. Biol. Med.* 1978; **159**: 201–203.

Provost, P.J. and Hilleman, M.R. Proc. Soc. *Exp. Biol. Med.* 1979; **160**: 213–221.

Siegl, G., Witz, M. and Kronauer, G. *Intervirology* 1984; **22**: 218–226.

Siegl, G., de Chastonay, J. and Kronauer, G. *J. Virol.* Methods 1984; **9**: 53–67.

Siegl, G. *In Viral Hepatitis* (ed.) Z uckerman, A. and Thomas, H.C. 1993 Churchill Livingstone 1993 p. 21.

Schultheiss, T., Kusov, Y.Y. and Gauss-Muller, V. *Virology* 1994; **198**: 275–281.

Schultheiss, T., Sommergrube, W., Kusov, Y. and Gauss-Muller, V. *J. Virol.* 1995; **69**: 1727–1733.

Tesar, M., Pak, I. and Jia, X-Y. *et al. Virology* 1994; **198**: 524–533.

Ticehurst, J.R., Ricaniello,V.R., Baroudy, B.M. *et al. PNAS* 1983; **80**: 5885–5889.

PART 2

HEPATITIS B VIRUS

History

Although serum hepatitis was recognized as a separate disease from infectious hepatitis, it was only in 1965 that Blumberg and colleagues first reported the presence of a serum marker, the *Australia antigen*, later to be associated with the agent responsible for this form of hepatitis. The virion was first seen in the electron microscope by Dane in 1970, hence the Dane particle, who clearly described its structure and correctly interpreted the nature of the associated smaller particles and tubular forms (Dane *et al.*, 1970). Soon after, the core of the virion was identified in infected hepatocytes and also isolated from full virions following detergent removal of the envelope (Almeida *et al.*, 1971).

Major advances during the early seventies, to which Robinson's group at Stanford was the main contributor (Robinson *et al.*, 1974a; Robinson *et al.*, 1974b), led to the characterization of the virus genome, virion-associated proteins (including DNA polymerase) and major antigen-antibody systems. Such information made possible the study of the natural history of the disease and the identification of the link between chronic HBV infection and hepatocellular carcinoma.

By the end of the seventies, the unavailability of a cell culture system for propagation of HBV remained the major impediment to the study of molecular events during infection. It was only with the development of genetic engineering in the early eighties that a spectacular change occurred in this area. As techniques for cloning and amplification of DNA became available, the virus genome was sequenced and putative genes identified. In 1982, the elegant experiments of Summers and Mason characterized the mechanism of replication and opened up a new phase in HBV research (Summers and Mason 1982). The pioneering work on the development of the vaccine by Smuzness around the same time must also be seen as a major contribution to HBV research (Szmuness *et al.*, 1980).

The Virus

Classification: The Hepadnaviridae

In view of the lack of genomic homology with other known viruses, HBV became the prototype of a new family, the *Hepadnaviridae*, which was soon to acquire other members. These now include the hepatitis virus of the woodchuck (*Marmota monax*) or WHV (Summers *et al.*, 1978), of the Peking duck or DHV (Mason *et al.*, 1980), of the ground squirrel or GSHV (Marion *et al.*, 1980) and of the heron or HHVH (Sprengel *et al.*). The virus also shows striking similarities in genetic organization and replication strategy with the Cauliflower Mosaic Virus (CMV).

The Virion

The complete virus particle is spherical, of 42–47 nm in diameter, and consists of a 25–27 nm diameter core surrounded by a lipoprotein envelope (Fig. 4). In addition small spheres and tubular forms of about 20 nm diameter are invariably present, sometimes reaching titres of 10^{13}/ml, and far outnumber the Dane particles (10,000–100,000:1). These structures are exclusively made up of envelope material (surface antigen or HBsAg) and are therefore non infectious.

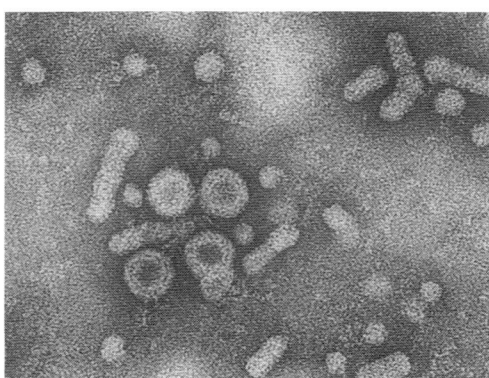

Fig. 4   Electron micrograph of HBV. Magnification: × 140,000. (original obtained and given to the author by the late Dr. R. Bird).

Density of virions and HBs particles is between 1.19–1.24 g/ml in CsCl and virions can be partially purified from other forms because of their higher density.

Electron cryomicroscopy with computer tri-dimensional reconstitution has shown self-assembled recombinant core particles to have icosahedral T4 symmetry with 240 constitutive subunits but with a less abundant population showing T3 symmetry and 180 subunits (Crowther *et al.*, 1994). The proposed structures contain holes or channels of up to 10 nm diameter, previously described in other transcriptionally active viral cores, which are thought to allow the entry of small molecules (Fig. 5). An inner shell of increased density has been demonstrated, with peaks mapping according to icosahedral symmetry, presumed to correspond to the packaged nucleic acid with some weak connections to the the core shell. These probably represent the arginine rich carboxy-terminal tails of the core protein. The structure of the core particle has recently been resolved further by electron cryomicroscopy to reveal spikes on the surface of the shell made up of dimers of core polypeptide in the form

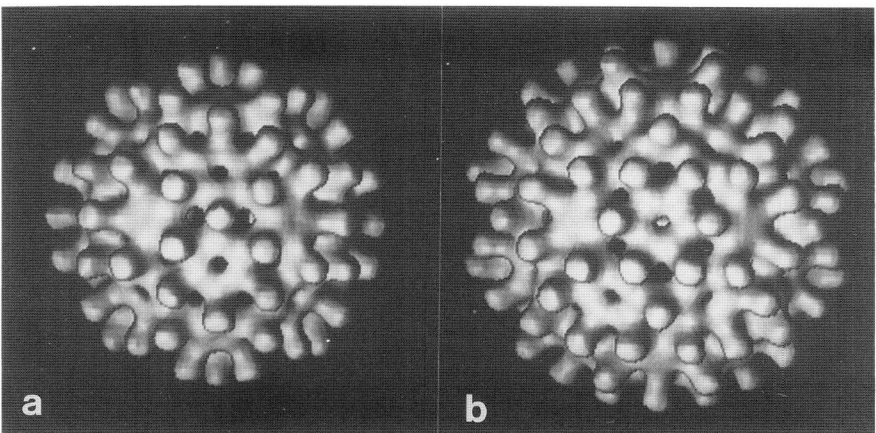

Fig. 5 Surface representations of the computer-reconstituted three dimensional maps of empty recombinant HBV core particles. View along a 2-fold axis of symmetry. a) small particle shows 90 protruding spikes arranged with dimer cluster T = 3 icosahedral symmetry; b) large particle shows 120 protruding spikes arranged with dimer cluster T = 4 icosahedral symmetry. The shells are penetrated by channels (From Crowther *et al.*, 1994).

of radial bundles of four α-helices (Bottcher *et al.*, 1997; Conway *et al.*, 1997). The virion also contains the DNA polymerase/Reverse Transcriptase enzyme. The molecular architecture of the whole particle has yet to be elucidated.

Virions from other hepadnaviruses have a similar morphology to HBV. They are also a minority species amongst a vast excess of small surface antigen particles and tubules. This is illustrated in Figure 6 which shows the structures associated with WHV.

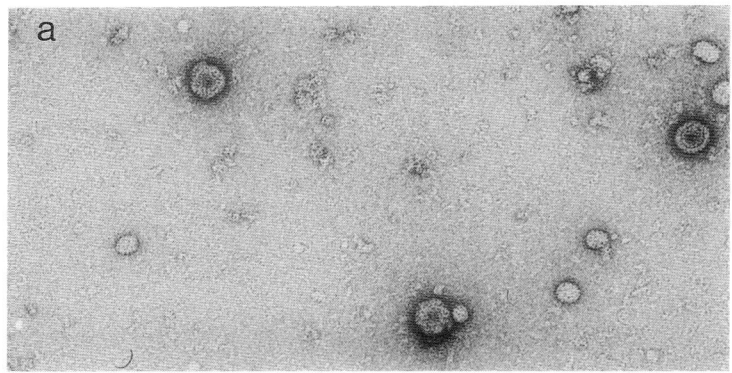

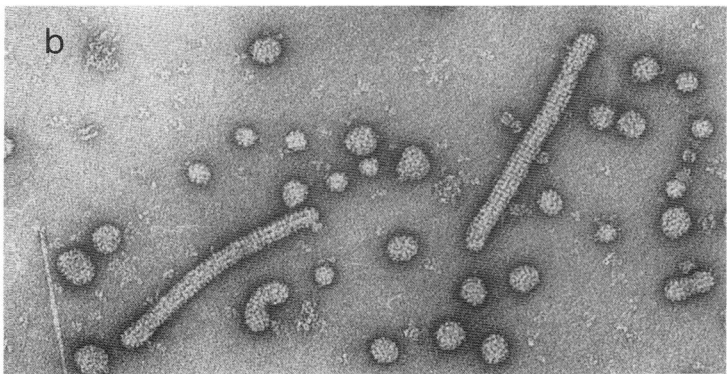

Fig. 6   Electron micrograph of Woodchuck Hepatitis virus (WHV) purified by equilibrium density ultracentifugation by the author. a) fraction enriched in full virions, magnification: × 107,000. b) fraction enriched in small spheres and tubules of WHsAg, magnification: × 175,000 (courtesy of Drs. B. Bottcher and A. Crowther).

The Genome

The genome is a circular partly double-stranded molecule of DNA consisting of one complete strand about 3,200 nucleotides long (minus strand) with a single, unique interruption and the complementary strand (or positive strand) which has variable length (Robinson *et al.*, 1974b; Summers *et al.*, 1975; Hruska *et al.*, 1977; Landers *et al.*, 1977) (Fig. 7). With a common 5′ end and different 3′ termini, the overall lengths of positive strands amongst different molecules

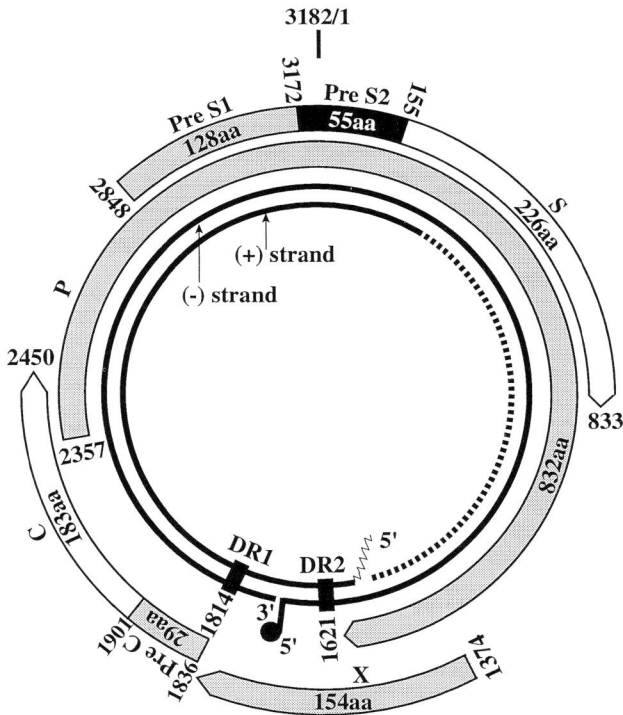

Fig. 7   HBV genome and Open Reading Frames (ORFs). Figures refer to nucleotide numbers. DR1 and DR2: Direct Repeats. 3182/1: Conventional origin of genomic map. ●: Terminal protein primer linked to 5′ end of minus strand; ᴟᴟᴟ: oligoribonucleotide primer covalently linked to 5′ of plus strand. PreS1, PreS2, s, PreC, C, P and X: HBV ORFs. ▬▬▬: DNA; ∎∎∎∎∎ incomplete plus strand.

range between 50% and 70% of the full size. A base-paired region of about 200 basepairs between the 5' termini ensures the circular configuration of the genome (Fig. 7). Two 11 base-pair repeats, present in all HBV genomes, have been designated DR1 and DR2 and are essential features in the virus replication strategy (see below). Analysis of the genome shows four major Open Reading Frames (ORFs) on different reading registers which correspond to the Pre-Surface/Surface, the Pre-Core/Core, P and X ORFs and account for one and a half times the information content of the full genome in a single reading frame (Fig. 7). This double usage of genetic information, read in two reading frames in some segments of the genome, imposes significant constraints on genetic variation of individual genes.

Correlation between viral protein amino acid sequence data and predicted ORFs has shown that there are three co-carboxy terminal proteins at position 833 encoded within the pres/S region. One, preS1, initiates at the AUT in position 2848 (for ayw3 subtype) and is 409 amino acids long; the next, preS2, initiates at the AUT at position 3172 and is 281 amino acids long and finally the major or small (s) protein initiates at position 155 and is 226 amino acids long (Fig. 7). A similar alternative use of initiation codons is seen in the core ORF where the core protein initiates at the AUT at position 1901 and terminates at position 2450, but may be preceded by a 29 amino acid precore sequence in the same reading frame starting from AUT at position 1814 (Fig. 7).

Of the two other reading frames, one P extends over more than three quarters of the genome from the AUT at position 2357 to the termination codon at position 1621 and is 832 amino acids long and finally the X ORF which initiates at position 1374 and terminates at position 1836 encodes a protein 154 amino acids long (Fig. 7).

In other hepadnaviruses genomes and genetic organization are similar to that observed in HBV. WHV and GSHV show greater homology between themselves (around 83.6%) and with HBV (60%–70% and 46% respectively). These three viruses now constitute the *Orthohepadnavirus* subgroup, whereas DHV and HHV, which show greater homology between themselves (79.4%) and less homology with HBV (<40%), WHV and GSH constitutes the *Avianhepadnavirus* sub-group. Other important differences between the groups include the absence of both the X and the preS2 genes in the avian viruses.

Virus Subtypes

HBV can be classified into subtypes that were originally defined by antibodies against HBsAg (serotypes). Of the defined determinants, one is common to all subtypes: *a*, and two pairs: *d* or *y*, *w* or *r*, were also commonly found producing subtypes *adw* or *adr* and *ayw* or *ayr* which have been further subdivided into nine subtypes *ayw1, ayw2, ayw3, ayw4, ayr, adw2, adw4, adrq⁺ and adrq⁻*. Preferential geographical distributions show that *adw* and *ayw* are more commonly found in Europe, North America and Africa and *adr* in the Far East (China, Japan, Korea, etc.). Individual amino acid changes have now been related to specific subtypes, *d* with a lysine at position 122 whereas *y* has an arginine; determinant *w* with a lysine at position 160 whereas *r* has an arginine. Because HBV genomes from a common subtype could be quite heterologous, subtyping has recently been carried out by sequencing the sAg gene.

Based on this information six different genotypes have been identified, designated A–F, differing by more than 8% in protein sequence (Okamoto *et al.*, 1988; Norder *et al.*, 1992). These include genotype specific amino acids that were not identified as part of serotype determinants and both *w* and *d* determinants which were seen to segregate into four of the new genotypes.

HBV Structural Proteins

Envelope Proteins

The particle is made of an envelope and a core. The envelope consists of three types of proteins or surface antigens known as preS1, preS2 and *s* (or major) (Fig. 7). Both preS1 and s can be found either unmodified or glycosylated whereas preS2 consists of two forms glycosylated to different extents (Heermann 1984). The molecular weights for these various HBV envelope constituents are 24 and 27 kD for the small or major surface antigen *s* (unglycosylated p24 and glycosylated gp 27 respectively); 31 and 33 kD for preS2 or M (glycosylated forms gp31 and gp33 respectively); and 37 and 39 kD for preS1 or L (unglycosylated p37 and glycosylated gp39 respectively).

The small surface antigen (or *s*) polypeptides are the most abundant forms in all three types of particles, i.e. Danes, 20 nm particles and tubules, whereas large surface antigen polypeptides (L) are relatively more abundant in Danes followed by the tubules and least abundant in the 20 nm particles. Medium surface antigen polypeptides on the other hand are not abundant in any of the three particles but are relatively more abundant than L-sAg in the 20 nm particles.

The current view of the major surface antigen transmembrane topology both in the virion envelope and in the ER membrane is shown in Fig. 8. The model proposes that the molecule contains four transmembrane domains of which only two have been shown to be involved in membrane interactions (Eble *et al.*, 1987; Bruss and Ganem 1991)and that termini, amino and carboxyl, lie in the lumen of the endoplasmic reticulum (Guerrero *et al.*, 1988). Particles can be assembled from major HBsAg alone in the absence of viral replication or production of nucleocapsids as seen after transformation of bacteria or transfection of cultured cells with HBsAg-expressing genetic constructs and also in HBsAg-producing hepatoma cell lines with no replicating HBV.

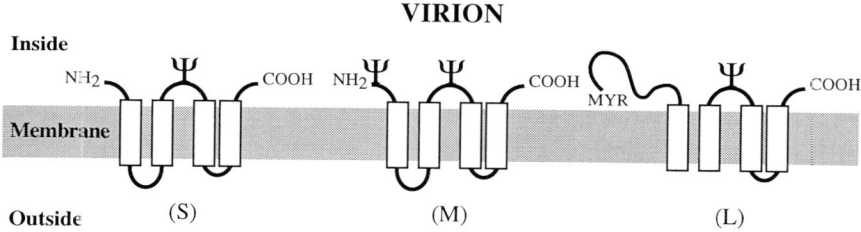

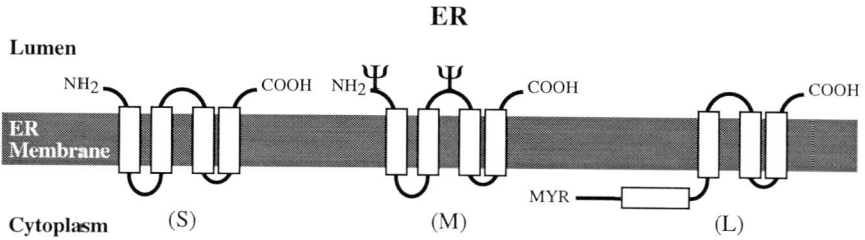

Fig. 8   Virion and Endoplasmic Reticulum (ER) topology of HBV envelope glycoproteins. S: small; M: medium; L: large; MYR: myristylation site; Ψ: glycosylation site. White boxes: Putative transmembrane domains.

Full virions also contain the two other surface antigen polypeptides, L which contains preS1 and preS2 in addition to *s* and M which contains preS2 and *s* (Fig. 7). Of these, L protein is essential for virion assembly whereas the contribution of M protein is still disputed, and both are displayed on the outside of the particles as shown by the binding of preS-specific monoclonal antibodies. In line with previous understanding of membrane topology M envelope protein has been shown to be translocated into the lumen of the endoplasmic reticulum during the process of biosynthesis. By contrast, in the case of L envelope protein, recent reports demonstrated that both the preS1-specific and the preS2 sequences are not translocated into the lumen of the reticulum and remain predominantly or exclusively cytoplasmic (Ostapchuk *et al.*, 1994; Prange and Steeck 1995) (Fig. 7). These findings explain why one site which is glycosylated in the M protein (in preS2) is not glycosylated in the L protein either *in vitro* or in virions (Fig. 8). Another modification of the L protein, i.e. the cytosolic addition of an N-terminal mirystic acid residue, is also well in agreement with the postulated initial transmembrane orientation of the L protein. Furthermore the present model also proposes that the preS sequences are involved in the cytoplasmic nucleocapsid interaction that triggers budding and sequences close to the preS1/preS2 boundary have been shown to be involved. Finally since pre S1 sequences are ultimately displayed on the outside of the full virions, a mechanism must operate for the 'extrusion' of such sequences, which lie initially in the cytosol. Although it has been advanced that this topological change may be mediated by the reorganization of the lipid bilayer the mechanism remains largely unknown.

In spite of the precise interactions required between surface proteins during assembly of the virions, these can nonetheless mix phenotypically with each other when derived from members of the same subgroup (HBV and WHV) although not when originating from different subgroups, i.e. HBV (*Orthohepadnavirus*) and DHV (*Avihepadnavirus*).

Core and *e* Proteins

The major core polypeptide is 183 amino acids long, has a MW of 21 kD and can dimerize and self-assemble in the cytoplasm into viral capsids. This can also be mimicked with recombinant core proteins expressed in *E.Coli*. The

C-terminus of the core polypeptide has an arginine rich-region starting at amino acid 150 which has been shown to have a nucleic acid binding function but to be dispensable for core particle self-assembly. Its presence, however, enhances encapsidation of nucleic acid and contributes to the stability of the capsids through protein nucleic acid interactions. Core particle architecture has been further elucidated by electron cryomicroscopy and computer reconstitution (above under VIRION).

Three repeated SPRRR motifs have been identified (Roossnck and Siddiqui, 1987; Schlicht *et al.*, 1989a; Yu and J. Summers, 1994) and its serine residues appear to be the major phosphorylation acceptor sites in the core protein sequence and to overlap the nuclear localization signal (Yeh *et al.*, 1990; Eckardt *et al.*, 1991). Inhibition of phosphorylation is accompanied by nuclear localization of core antigen and also appears to be involved during the cell cycle as the S phase is accompanied by predominantly cytoplasmic accumulation and nuclear localization is observed during G1 or the confluent quiescent G0/G1 phase (Liao and Ou, 1995). Transposition of these findings to the situation *in vivo* may shed light into the significance of different intracellular core Ag staining patterns, i.e. nuclear and cytoplasmic, which appear to be related to the disease state during natural infection (Chu and Liao, 1987).

A secreted core-related antigen, *e* antigen, has been known for a long time and although shown to be essential for viral replication its function remains unknown. It is derived from the precore polypeptide which is translated from the pre-C RNA (see below) and initiates at the AUG codon at position 1814 (Fig. 7) thus comprising both precore (containing a signal sequence) and core ORFs. After synthesis, the polypeptide is directed to the ER by the signal peptide in the precore region. Cleavage of the signal sequence in the lumen of the ER is followed by further proteolytic processing by a host protease at amino acid position 149, just upstream of the arginine-rich C-terminus, and ultimately by secretion of a soluble 17 Kd polypetide (*e* antigen) into the circulation. Defective processing in the ER after the initial signal peptide cleavage can occur and is followed by the appearance of *e* antigen in the cytosol and also in the nucleus.

HBV Non-Structural Proteins

DNA Polymerase/RNA Transcriptase

The polymerase is the product of the P ORF (Fig. 7) and is central to the replicative strategy of HBV. It is a protein of about 84,000 MW which recognizes its own pregenomic mRNA, i.e., it is capable of acting in *cis*. It interacts with a specific sequence (*epsilom*) of pregenomic RNA (see Replication below) and with core protein to package pregenomic RNA molecules into immature cores and will also package other heterologous-bearing RNAs. After completing a round of synthesis, the polymerase molecule does not appear to disengage from the endogenous template as shown by its inability to transcribe added templates, *in vitro,* in detergent-treated permeabilized virions (Radziwill *et al.*, 1988).

The polymerase is a multifunctional protein showing amino acid homology with DNA polymerase RNAse H domains of reverse transcriptases of retroviruses (Toh *et al.*, 1983; Khudyakov and Makhov, 1989; Radziwill *et al.*, 1990). Its different domains include (from the amino terminus) a terminal protein, a spacer region, the polymerase/Reverse Transcriptase and the RNAseH domain which lies at the carboxyl end (Fig. 7). The protein appears not to be processed and each individual function involves the appropriate catalytic or DNA initiation domain which is recruited and operates in the presence of the others. Although the DNA polymerase function can be easily demonstrated in serum-derived detergent-treated Dane particles (Robinson, 1974b), attempts at extracting and purifying the enzyme from infected serum or liver did not succeed and failure to engage added templates indicated that the enzyme remained tightly bound to the original template (Radziwill *et al.*, 1990).

Low abundance of the virion-associated and recombinant polymerase prevented for some time its identification and sizing by conventional immunoblotting or activity gel analysis. The problem was eventually overcome by the introduction of a phosphorylation site into recombinant polymerase which, once labelled with ^{32}P, revealed the full length protein inside the virions (Bartenschlager *et al.*, 1991).

Recently, both *in vitro* translation and expression of recombinant polymerase have met with some success, first in the hepadnavirus of ducks, DHV, and then in HBV (Wang and Seeger, 1992; Tavis and Ganem, 1993). This led to the synthesis of a molecule with RT and terminal protein activities if devoid of RNAseH function. In a more recent but still unconfirmed report, a 70 kD protein with polymerase and RT activities on exogenous templates has been solubilized from serum-derived HBV cores and shown to retain enzymatic activity after chromatographic purification(Shin and Rho, 1995).

X Protein

The product of the X gene is a protein of 154 amino acids which has yet to be characterized from naturally infected hosts. Its presence has so far only been documented by staining with antibodies raised against synthetic peptides or indirectly, by the presence of antibodies in infected serum. The protein appears to be a 'recent' addition to the genome based on the distinct mammalian type codon-usage and on the fact that it is absent in the avian members of the hepadnavirus family. *In vitro* it has a transactivator function which has been demonstrated both in homologous and heterologous systems (Twu and Schloener, 1987; Seto *et al.*, 1988; Spandau and Lee, 1988) and *in vivo* it has been shown to be essential to establish the infection of woodchucks with WHV (Zoulim *et al.*, 1994). Since the last 14 amino acids are not conserved amongst different hepadnaviruses it has been suggested that the last seven conserved residues, 134–140 (Leu-Gly-Gly-Cys-Arg-His-Lys) may be essential for the transactivation function of the X protein. This prediction was supported by the finding that aas 132–139 (especially Phe-132, Cys-137 and His-139) were found to be essential for the transactivation function tested in a chloramphenicol acetyltransferase assay. Two other regions (aas 45–52 and 61–69) were also found to be required and all three are conserved amongst hepadnavirus X proteins. Neither aas 5–27 (Arii *et al.*, 1992) nor the 12 carboxy-terminal residues (Takada and Koike, 1990) were found to be essential for transactivation.

The protein appears to be produced in small amounts and to be weakly immunogenic although anti-HBx antibodies have been reported in HBV carriers and in cases of HBV-associated hepatocellular carcinoma.

The intracellular distribution of the X protein has been the object of conflicting reports, describing its accumulation in the cytoplasm (Katayama *et al.*, 1989), nucleus and cytoplasm (Hu *et al.*, 1990), or around the nucleus

(Chisaka *et al.*, 1987), but recent and more detailed investigations have demonstrated that in transfected cells the protein accumulates predominantly in the cytoplasm, un-associated with membranes, vesicles or identifiable structures, and also that there is a definite, if minor, fraction in the nucleus (Doria *et al.*, 1995).

Studies on the molecular basis of the transactivating properties of the X protein have generated many diverse observations and almost as many mechanisms have been suggested. These range from a direct effect on the transcription machinery in the nucleus, RNA polymerase and transcription factors (Seto *et al.*, 1990; Cheong *et al.*, 1995; Qadri *et al.*, 1995) to the stimulatory effect on cytoplasmic *Ras*-GTP complex formation, followed by rapid induction of a signalling cascade linking *Ras*, *Raf* and MAP kinases and subsequent transcriptional transactivation via NF-kB and AP-1 (Benn and Schneider, 1994; Doria *et al.*, 1995). Other proposed mechanisms involve interactions with a serine protease, the proteosome complex (Huang *et al.*, 1996), p53 (Feitelson *et al.*, 1993) and also the promotion of a transcription function that activates enhancer EI. Whilst further clarification is required on the underlying molecular basis for X protein-mediated transactivation, one finding appears to be well established: that the X protein does not interact directly with DNA and that its effects are likely to be mediated by other proteins, viral or host.

Molecular Biology of HBV Replication

Transcription and Translation of HBV Genes

The enzyme that transcribes HBV DNA is presumed to be RNA Polymerase II and viral transcription in infected cell extracts has been shown to be inhibited by the inhibitor α-amanitin. Four different promoters controlling the synthesis of an equal number of transcripts have been identified. The more abundant mRNA species is about 2.1 kb long (Fig. 9) and is the template for the synthesis of either pre-S2 (or M) protein or small antigen *s*. Heterogeneity at the 5' end accounts for different transcripts which become templates for one or the other form of the antigen. Preferred synthesis of sAg, even from mRNAs bearing

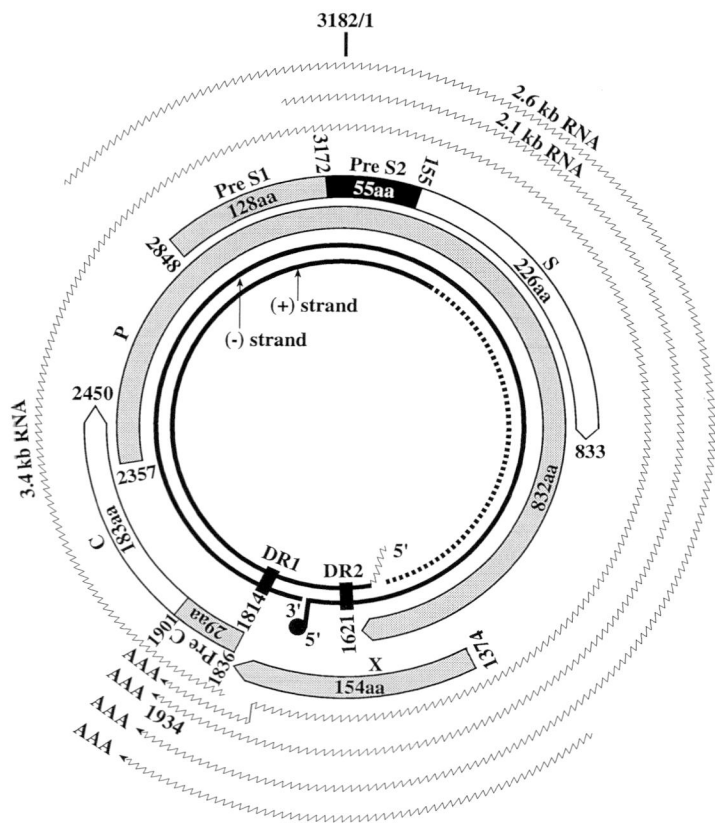

Fig. 9 Transcription map of HBV. ∿∿∿: Transcripts with sizes are shown; AAA: polyA tail. Other details as in Fig. 6.

the preS2 initiation codon, reflects the more favourable sequence context for ribosome initiation at the second AUG codon. Such heterogeneity of the 5′ termini of the 2.1 kb transcript probably reflects the absence of a consensus TATAA sequence upstream of the transcription initiation sites. Instead a 'heterodox' promoter has been identified which shares limited homology with the late promoter of SV40 (Cattaneo *et al.*, 1983). The promoter has been shown to be active in a number of cell types after transfection but shows highest activity in hepatic cell lines and preferential hepatic expression in transgenic mice (Standring *et al.*, 1984; Dubois *et al.*, 1980; Araki *et al.*, 1989; Babinet *et al.*, 1985; Choo *et al.*, 1991). Such hepatic preference is thought to be mediated through viral enhancer II.

Termination of the transcript occurs some 20 nucleotides downstream from the sequence TATAAA, located within the coding region of the core gene, which appears to be the functional homolog of the consensus AATAAA similarly sited in other polII transcripts.

Transcripts of about 3.4 kb (Fig. 9) serve as mRNAs for both core and polymerase multi-functional protein , as pregenomic RNA template for reverse transcription during replication, and as mRNA for precore protein which is ultimately processed in the ER into *e* antigen. Heterogeneity of the 5′ termini of these transcripts was thought to account for the inclusion in some, but not in others of the precore initiation codon at position 1814. This would account for the synthesis of both transcripts, and a stretch of 15 nucleotides (nt 1790– 1804) was reported to be sufficient for their correct initiation at nucleotides 1794 and 1823 respectively (Chen *et al.*, 1995). Recent evidence however appears to support the existence of two distinct promoters: one for pregenomic RNA from which core protein and polymerase are translated and which also functions as a replicative intermediate (pregenomic RNA), and another for pre-C RNA, from which precore mRNA is translated (Yu and Mertz, 1996). The pregenomic RNA promoter has been identified by genetic analysis with the predictably positioned TATA box-like sequence 5′-TTAAA-3′ (1760–1764 for *adr* type) and *Inr* element and that of pregenomic RNA with the sequence 5′-CATAAAT-3′ (1790–1796) and an *Inr* element. High-resolution mapping shows that pre-C RNAs are much more heterogeneous at the 5′ end than pregenomic RNA. The promoters partly overlap but can be separated and appear to be separately regulated by tran-acting host factors: *Sp1* specifically increases up-regulation of the pregenomic promoter but not the pre-C, and HNF4 specifically down-regulates the pre-C (Yu and Mertz, 1996). Differential regulation in synthesis of the two HBV RNAs may be important during the virus life cycle but its significance will require a better understanding of the role of *e* antigen, the ultimate product of the translation of pre-C mRNA. That there can be a functional dissociation between the transcription of the two RNAs is indicated by the finding of HBV *e* antigen negative patients with a mutation in the precore promoter, unchanged (wild-type) precore/core ORF and actively replicating virus (Okamoto *et al.*, 1994; Takahashi *et al.*, 1995). Alternatively, such differential transcription (Chen *et al.*, 1995) may be the result of modulation by different transcription factors on the same promoter as illustrated by the observation that a liver-enriched transcription factor which binds to nts 1759–1769 appears to be solely involved with precore RNA synthesis (Buckold *et al.*, 1996).

When transcribing the 3.4 kb long RNA, the polymerase ignores the termination site during the first passage, but recognizes it during its second passage thus generating a transcript which is longer than the full length genome (3.4 kb rather than 3.2 kb) (Fig. 9). The molecular basis for the termination 'leakage' is not established but is thought to involve the obligatory co-operation of other DNA sequences upstream of the promoter in order to generate an effective termination signal (Cherrington *et al.*, 1992). Initiation of transcription with binding of the polymerase to the promoter may also interfere with the formation of such a termination complex which only occurs during the second passage.

From the single transcript, both core and polymerase are synthesized. Since the two proteins are encoded on two overlapping ORFs it was originally thought that initiation was common to both proteins, the core AUG, with core being translated normally and polymerase as part of a fusion protein of truncated core and polymerase as found in retroviruses. This would arise from a ribosomal frameshift (the P frame is +1 in relation to the core frame) and the fusion protein would be ultimately processed by proteolytic cleavage to generate the polymerase molecule. This prediction was to prove incorrect, both for DHV and for HBV, in the absence of demonstrable core/protease fusion protein and after evidence was obtained to demonstrate ribosome initiation from the internal AUG at position (Chang *et al.*, 1989; Schlicht *et al.*, 1989; Jean-Jean *et al.*, 1989; Roychoudhury and Shih, 1990). Although the mechanism for the initiation of P ORF translation remains unclear, the currently accepted model predicts that ribosomes scan the mRNA from the capped 5′ terminus before reaching the internal AUG (Fouillot *et al.*, 1993).

The third transcript, or 2.6 Kb (Fig. 9) is the template for preS1 (L surface antigen) protein (Fig. 7). It is a transcript of very low abundance in naturally infected tissues, transcribed weakly after transfection into differentiated hepatoma cell lines and almost absent in non-hepatic cell lines. This appears to be explained by the dependence of the preS1 promoter upon hepatocyte nuclear factor 1 (HNF-1)(Chang *et al.*, 1989). The normally weak (or downregulated) promoter contains the consensus TATA box. At the 3' end the 2.6 kb transcript is co-terminal with both the 2.1 kb and 3.4 kb transcripts (Fig. 9).

Transcription of the X gene is less well understood. Although the gene is copied in all other three transcripts (2.1, 2.6 and 3.4 kD), translation from an internal promoter is normally expected to be very inefficient. It is currently assumed that translation of the X protein occurs from a specific 0.8 kb mRNA in spite of a poor sequence context for initiation from the initial ATG. A region with promoter activity has been identified (Nakamura and Koike, 1992) and the major start sites for the X gene transcripts at 1117 and 1125 are at the downstream boundary of enhancer I (Yaginuma *et al.*, 1996)

Two enhancer elements, enhancer I (*EnI*) and enhancer II(*EnII*) have been described in the HBV genome. Both have an upregulating effect on heterologous promoters which is independent of position and orientation and which is significantly more marked in hepatocytes or differentiated hepatoma cells. In fact *En II* is totally dependent on hepatocytic cells for activity (Yee 1989; Wang *et al.*, 1990; Yen, 1993; Yuh and Ting, 1993). Each is located 5' to one promoter, *EnI* to the X promoter and *EnII* to the genomic RNA promoter in the X gene. *EnI* is located between the surface and the X genes and is about 200 nucleotides long. It has been shown to upregulate all the HBV promoters in transfection experiments using both reporter and HBV genes (Antonucci and Rutter, 1989; Hu and Siddiqui, 1991). Many factors, some hepatocyte-specific, C/EBP, HNF-4, HBLF) others not (NF-1, AP-1, NFkB and EF-C), have been shown to bind to this sequence (Trujillo *et al.*, 1991; Garcia *et al.*, 1993; Guo *et al.*, 1993). *EnII* has a particularly significant stimulatory effect upon the genomic RNA promoter but is also active on the S promoter, although not on the preS1 promoter, and the stimulatory effects of two cellular factors abundant in hepatocytes (C/EBP and HNF-4) appear to be mediated by this enhancer.

More detailed mapping of the individual binding sites for the various factors and their functional significance could only be carried out in the duck hepatitis virus. Results from such experiments have demonstrated both a certain degree of functional redundancy and also specific effects relating to aspects of transcription control.

Cell Attachment and Entry

The mechanism by which HBV enters the hepatocyte is poorly understood. The ability to infect hepatocytes of young ducklings with DHV, but not of human and woodchuck hepatocytes with HBV or WHV, has meant that most of the information on attachment and internalization has been derived from the avian hepadnavirus.

Attachment of the virus can occur at low temperature but does not lead to infection and only by raising the temperature to 37° does the infection proceed. Binding of the virus involves two components, one of low-affinity and non-saturable and another of high affinity and saturable (Klingmuller and Schaller, 1993). Envelope viruses can be internalized by an endocytotic pathway or as a result of membrane fusion. In the first instance acidification of the interior of the endocytotic vesicle is thought to cause a conformational change in the viral envelope which leads to fusion with the endocytotic membrane and extrusion into the cytosol of the vesicular contents. Lysosomotropic agents like ammonium chloride, by preventing acidification of the vesicles' interior can be used to determine whether such a mechanism is involved. Although results of such experiments with DHV were initially discordant (Offensperger *et al.*, 1991; Rigg and Schalle, 1992), a recent report appears to show that the operating mechanism leading to internalization of DHV is both pH-independent and energy-dependent, thus, implicating an endocytotic pathway but a diverse uncoating mechanism from the low-pH driven fusion described for other viruses (Kock *et al.*, 1996).

HBV attachment to the hepatocyte appears to be mediated by a specific interaction between a receptor on the envelope of the virion which has been mapped to preS1 (Neurath *et al.*, 1986; Pontisso *et al.*, 1989) and a cell receptor which has so far proved elusive.

Mapping of the virus receptor to the preS1 region was suggested by *in vitro* peptide binding and antibody blocking experiments involving both HepG2 membrane preparations and whole cells and further supported by blocking DHV infection with preS-containing protein preparations but not with *s* alone (Klingmuller and Schaller, 1993).

In what concerns the cellular receptor a variety of different molecules have been proposed. These include amongst others, serum protein apolipoprotein H (Medi *et al.*, 1994) and endonexin 2 (Hertogs *et al.*, 1993) for HBV and a 180 kD surface glycoprotein for DHV (Kuroki *et al.*, 1994). The latter binds specifically to pre-S1 and is blocked by neutralizing antibodies. These observations strongly suggest a role for that protein in virus entry. Against such function, however, is its ubiquitous distribution, including a variety of cells known not to be infected with HBV. The ultimate test, i.e., the ability to confer infectibility upon its transfection into cells previously non-infectable with the particular virus has yet to be met by any of the many receptor candidates.

All subsequent intracellular steps leading to the delivery of HBV DNA to the nucleus remain poorly understood. Whether capsids once released enter the nucleus through the nucleus pores, directed by putative nuclear targetting sequences found in the carboxy terminus of the core protein, or simply release the DNA into the nucleus is not known. Once the DNA gets into the nucleus, removal of the RNA primer attached to the positive strand and of the protein primer attached to the minus strand (see below) as well as DNA ligation into covalent closed circular templates must take place before replication can start. All these steps appear to be carried out efficiently by the host cell machinery.

Replication of HBV Genome

The replication mechanism adopted by HBV appears to be unique to hepadnaviruses and to cauliflower mosaic virus. The molecular events occurring during replication were first elucidated for duck hepatitis virus by the elegant experiments of Summers and Mason in 1982 and demonstrated for HBV (Monjardino *et al.*, 1982; Fowler *et al.*, 1984). After infection and uncoating the genome migrates to the nucleus where the partially double-stranded molecule is first filled in presumably by the attached DNA polymerse/RT present inside the virion and then ligated into a supercoiled molecule by host ligase (Fig. 10). This covalently closed circular molecule (CCC) then becomes the template for transcription of the minus strand, putatively by host RNA polymerase II. The transcripts show heterogeneity at the 5′ termini possibly reflecting the absence of a TATA promoter motif or/and the existence of two

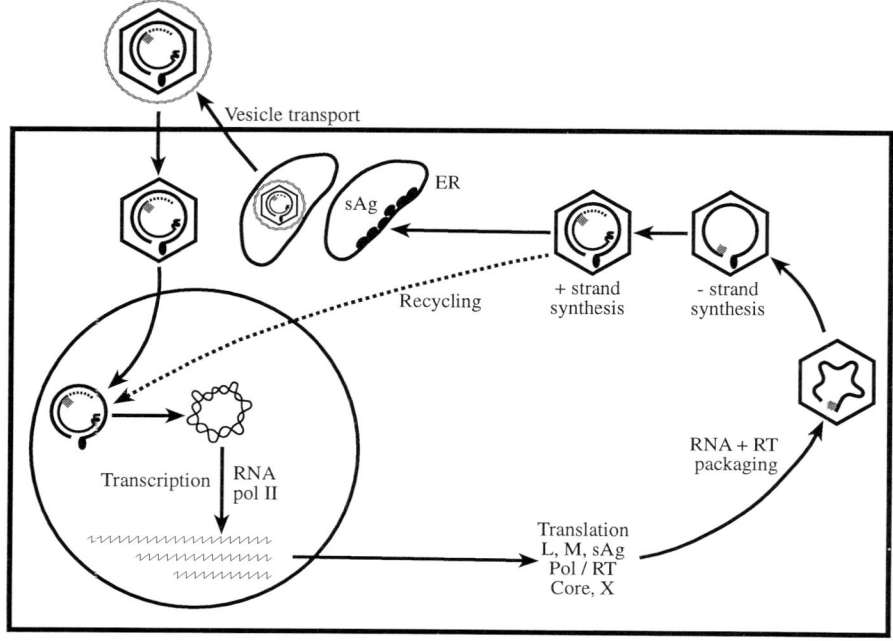

Fig. 10   HBV infection cycle. RNA Pol II: RNA polymerase II.

partly overlapping promoters (see HBV transcription). Transcription of the minus strand 'ignores' the termination signal during its first passage but recognizes it during the second passage, thus generating a 3.4 kb transcript, larger than full-genome length, which exits the nucleus (Fig. 10). These molecules containing the DR1 region at both ends will become templates for replication (pre-genomic RNA) and mRNAs for core and polymerase translation (Fig. 9).

Translation of the long 3.4 kb mRNA leads to the synthesis of RT/DNA polymerase and of core polypeptide structural units. The latter first dimerize, then self-assemble into immature cores, and subsequently internalize a complex which is formed between the pregenome RNA and RT. The way in which such a complex is generated has been recently elucidated by the groups of Ganem and Seeger (Wang and Seeger, 1993; Tavis *et al.*, 1994) and involves the recognition by the RT of a packaging signal, mapped to a stem loop structure at the 5′ end of the pregenomic RNA. Introduction of single base mutations or two-base mutations which preserve base-pairing in the stem has demonstrated that secondary rather than primary structure is all important, as are both the single stranded 'bulge' half-way up the stem and the terminal loop (Fig.11).

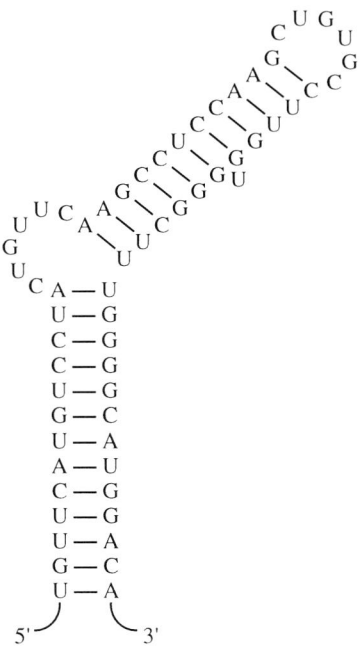

Fig. 11 The ∈ domain of HBV pregenomic RNA.

After recruitment into the immature cores, reverse transcription of pregenomic RNA is initiated on a terminal protein primer. That domain of the polymerase appears to interact with the pregenomic RNA at DR1 in such a way that the nascent DNA strand is initiated with a G residue which forms a covalent bond with a tyrosine residue of the protein primer. Since the pregenomic RNA contains two copies of DR1, one at each end, it was not clear which one was involved. Initial predictions pointed to the DR1 near the 3′ (where synthesis would have started) but this was confounded by the finding of a short four nucleotide oligo linked to the primer protein attached to the 5′ DR1 (Fig.12). It has since become clear that such a complex is subsequently transposed to the

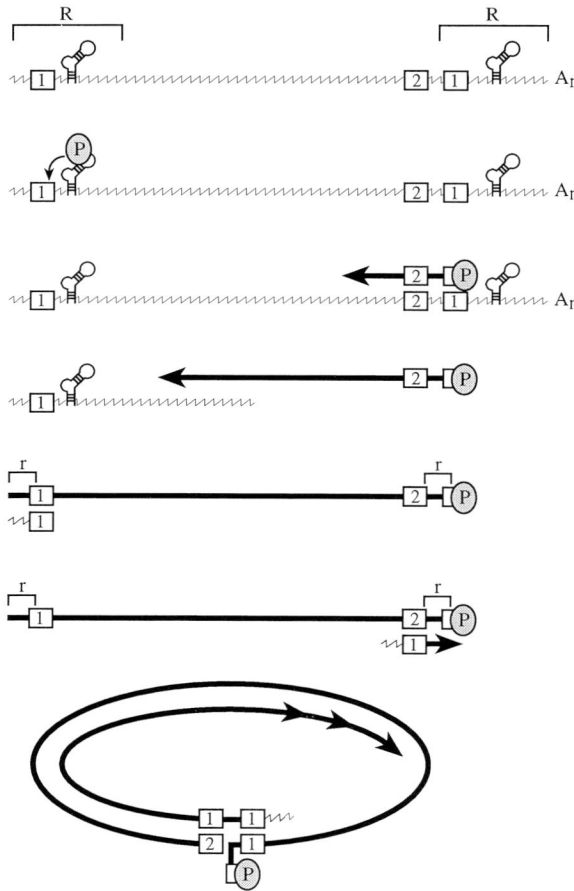

Fig. 12   HBV DNA replication. R: repeated sequence at either end; $\boxed{1}$ and $\boxed{2}$:direct repeats; P: polymerase. ⋀⋀⋀ : RNA; ▬▬ : DNA. Arrow indicates direction of synthesis. A_n: poly A tail at the 3′ end of pregenomic RNA.

3' terminal DR1 where it is re-positioned for subsequent synthesis to take place. Figure 11 illustrates these various stages. As reverse transcription of pre-genomic RNA progresses, removal of transcribed RNA is carried out *pari passu* by the associated RNAse H activity. Such digestion does not however proceed to the end of the RNA template and a 5'-terminal 18 nucleotides long oligoribonucleotide is retained which includes the DR1 and 6 nucleotides of unique terminal sequence. The size of this oligo has been shown to be sequence-independent and exclusively determined by the distance to the 5' terminus of the RNA. It's size is likely to correspond to the space between the leading polymerase and the trailing RNAse H domains (Loeb *et al.*, 1991). Such an oligoribonucleotide is then transposed to base-pair with the other DR1 located near the 5' end and becomes a primer for plus strand synthesis (Fig.12). Extension first proceeds by copying the minus strand to its 5' end. Progression of plus strand synthesis beyond this point requires a reconfiguration of the template which involves the dislodgement of the original DR1 (close to the 5' end of the template) by the DR1 at the 3' end (Fig.12). Once the latter is hydrogen-bonded to the growing plus strand, further extension of this strand can proceed unimpeded to the point where it terminates (which is short of full length and variable between different molecules). Failure to transpose the 5' oligoribonucleotide to prime the plus strand results in *in situ* priming and synthesis of linear double-stranded DNA molecules. Such molecules have been shown to account for as much as 7%–20% of the replicative intermediates and can be enveloped.

Although the reason for the heterogeneity of plus strand 3' termini is not known, it is possible that a change in core conformation occurs beyond a certain plus strand minimum length and that the final strand length may ultimately reflect small changes in elongation rate. As the core conformation changes, possibly completing the maturation process, a recognition signal may be revealed that will mediate the targetting of the core to the membrane-anchored envelope leading first to particle release into the lumen of the endoplasmic reticulum and ultimately to its exit from the cell via an exocytotic vesicle (Fig.10).

Envelopment of the capsids has been recently shown to depend on DNA synthesis, since RNA is not found in enveloped particles, and to show a strong preference for DNA molecules which are already partly double-stranded (Wey *et al.*, 1996) though not necessarily circularized (Yang and Summers, 1995).

The replication model as outlined above does not account for the amplification of the original infecting genome which is converted into a supercoiled molecule (CCC)and transcribed by polII. The presence of multiple copies of CCC DNA in the nucleus, which constitute a repository of HBV templates in the infected cell, cannot therefore be explained by the replication of the infecting genome (Tutleman *et al.*, 1986). Instead such molecules appear to be generated by the diversion from the ER to the nucleus of some newly-assembled core particles (Fig.10). How the molecules of partially double-stranded (relaxed circle) DNA reach the nucleus is presently unclear. In addition to this pathway mutated CCC DNA molecules of both duck and woodchuck hepatitis viruses with either deletions or insertions have recently been described and were shown to represent an alternative pathway resulting from *in situ* priming of plus strand (see above) with double-stranded linear molecules then becoming the substrate for non-homologous recombination (Yang and Summers, 1995; Yang *et al.*, 1996) . The significance of such a mechanism during natural infection is not known but the suggestion has been advanced that persistent HBV viral infection with low levels of viral replication may be associated with defective CCC HBV DNA molecules and that the non-homologous recombination mechanism outlined here may also underlie the integration of HBV DNA into hepatocyte genomic DNA (Yang *et al.*, 1996).

Growth of HBV in Cell Culture

Neither HBV nor the related viruses from the woodchuck (WHV) or the ground squirrel (GSHV) grow in cell culture. By contrast the duck virus (DHV) can infect primary hepatocytes derived from young ducklings and produce progeny virus which is released into the medium. Inability to infect cells in culture is likely to reflect the lack of a specific cellular receptor, still uncharacterized, to mediate virus entry into the cell.

The alternative approach for the study of HBV is to bypass the receptor by transfecting the naked viral genome directly into suitable hepatoma cells. A successful transfection depends on the degree of differentiation of the hepatoma cell line with Huh-7 and HepG2 human and LMH chicken hepatoma lines being the most effective. Such a system provides a means for analysing the stages of the virus cycle beyond uncoating but can only produce limited amounts of virus since no re-infection of the cells with progeny virions can occur. Alternatively, cell lines can be transfected for continuous virus production. This has been achieved after co-transfection with HBV DNA (either closed circles or plasmid containing two copies of genome) and an antibiotic-resistance marker followed by selection and cloning. In this case replication is initiated by transcription of HBV DNA copies stably integrated into the host cell genome and replication is then thought to proceed in the same way as from episomal DNA. An example is the HepG2-derived 2.2.15 cell line (Sells *et al.*, 1987) which has been extensively used in the testing of antivirals. Stable HBV transfected cell lines constitute to date the only system for continuous production of virions, albeit with rather modest yields.

Pathogenesis

The histological changes observed in the HBV-infected liver are not specific to this virus and consist of inflammation, with a predominance of lymphocytes, and hepatocyte degeneration and necrosis which, depending on severity, may eventually cause severe disruption of normal tissue organization. Viral markers are detected by immunostaining, HBsAg in the cytoplasm predominantly around the cell membrane, and core antigen in the nucleus.

HBV is normally non-cytopathic to hepatocytes as illustrated by a number of situations where active viral replication is associated with very minor or no cellular injury. Such examples include 'healthy' carriers of the virus, cases of persistent infection where in spite of a localized mild inflammatory infiltrate and active viral replication there is no histological evidence of cellular necrosis (Chu *et al.*, 1985) and transgenic mice expressing high levels of replicating virus where again no cell injury is observed (Guidotti *et al.*, 1995).

Current consensus on the mechanisms leading to the necro-inflammatory lesions of hepatitis is that they result from the cell-mediated immune clearance of infected hepatocytes and that progression to chronicity is the result of the inadequacy of such a response (Chisari and Ferrari, 1995). This view has received support from the induction of a histologically acute hepatitis-like condition in transgenic mice expressing HBsAg as a result of the adoptive transfer of CD8$^+$ MHC class I(L^d)-restricted HBsAg specific cytotoxic T lymphocytes (CTL) (Moryiama *et al.*, 1990). These studies also suggest that although the changes are initiated by CTLs, other cells, particularly macrophages and antigen non-specific cytokines, are subsequently recruited to the initial focus of inflammation and are ultimately responsible for most of the hepatocyte damage (Chisari and Ferrari, 1995).

Progression of HBV infection beyond the acute stage, i.e., to chronicity is the result of the inadequacy of the immune response. Whereas in the successfully resolved acute infection a strong, polyclonal, T-cell immune response against core, envelope and polymerase epitopes has been documented accompanied by effectively neutralizing anti-HBs circulating levels in those cases which become chronically infected, a weak and narrowly targetted T-cell response unaccompanied by the production of neutralizing antibodies appears to be the norm. If a defective immune response is thought to be the cause of persistence, its underlying cellular/molecular basis is not understood.

The observed high HBV carriage rate (>90%) in the offspring of HBV-infected mothers suggests that viral persistence is associated with tolerance occurring either as the result of transplacental infection or of transfer of subviral antigens or, alternatively, with the exposure of the neonate to virus when the immune system is relatively immature. However, successful immunization with HBsAg of newborns in general and of those born to mothers with actively replicating HBV in particular clearly demonstrates that in newborns the B cell immune response in respect to HBsAg is adequate and that HBsAg-specific T and B cells are retained and therefore not clonally deleted as a result of exposure to viral proteins during fetal life. If the acquisition of a permanent state of tolerance in this setting has not been demonstrated the possibility of it occurring during adult onset of infection is no better established. In this context neither high viral loads leading to CTL depletion nor CTL killing resulting from overstimulation (as described by superantigens) can be implicated.

Amongst other mechanisms leading to viral persistence the escape from T cell recognition as a result of high mutation rate in the sequence of a dominant epitope has been recognized as a possible mechanism. However, its contribution is unlikely to be significant during acute infection in the presence of a strong response targetted at a variety of epitopes in envelope, nucleocapsid and polymerase. In spite of reported natural mutations reducing HLA binding and T cell receptor recognition of a particular HLA-A2-restricted epitope in HBcAg in chronic HBV infections these constitute a very small proportion of cases and in most others no mutations were found in identified HLA A2-restricted CTL epitopes in HBV proteins.

What about the virus, could selection other than immune selection, generate variants capable of outstripping the immune response? Although theoretically possible mutations of the virus described so far (see HBV Mutants below) have failed to identify a virulence determinant(s) associated with increased levels of replication that could be correlated with either fulminant hepatitis or viral persistence.

The Disease

Natural history

The natural history of the disease has been well documented by assaying for viral markers and antibodies against viral antigens. Acute infection is characterized by the presence of direct markers of viral replication, i.e., viral DNA and viral DNA polymerase, of an indirect marker (*e* antigen) and of antibodies against core (anti-core). This is followed by the appearance of antibodies against *e* (anti-e), continuing decline and cessation of viraemia and detection of antibodies against surface antigen (anti-HBsAg). The latter is associated with the neutralization of the virus and resolution of infection as shown by the accompanying normalization of all other biochemical markers of hepatic dysfunction (billirubin, transaminases, etc).

In about 5% of the cases, the infection does not resolve and instead progresses to chronicity during which the virus persists for many years, associated with either no detectable clinical disease, mild disease or very active disease. Replacement of functional liver cells by connective tissue with regenerative nodules of hyperplastic liver cells (cirrhosis) ultimately follows the active inflammatory stage. This is characterized by major disruption of the liver architecture and marked hepatic dysfunction. Serious hemodynamic and hemostatic complications develop as a result of the increased resistance to hepatic vascular flow and reduced synthesis of plasma proteins.

A further complication of chronic HBV hepatitis is the development of hepatocellular carcinoma (see below).

In a particular group of patients where HBV is vertically transmitted (from mother to baby) the incidence of chronicity is much higher, reaching about 90%, an effect thought to be due to partial tolerance. The virus is nonetheless associated with liver injury and these patients normally evolve to cirrhosis and in a significant number of cases to hepatocellular carcinoma.

HBV Mutants

Current availability of genetic engineering techniques and the use of PCR amplification have greatly facilitated the cloning and sequencing of HBV DNA. As a result a multitude of studies have been published where correlations have been claimed between the course or the severity of the infection and the presence of a predominant mutated HBV DNA molecule. Most of these studies relate to a single time point observation, none have been independently confirmed in another system (cell culture or animal model) and only in few instances has the whole genomic sequence been analysed to exclude mutations in regions other than the one the authors *a priori* elected to study.

Of the various regions where mutations have been described the pre-core region was the first and is still one of the most commonly studied. Mutations are found in this region which are accompanied with premature termination and abolition of *e* antigen synthesis. The most commonly reported of these mutations appears to be the G → A at position 1896 (Carman *et al.*, 1989; Brunetto *et al.*, 1989). Such an occurrence appears to be part of natural seroconversion from eAg to anti-e in some regions of the world and to be associated in Mediterranean countries with chronic HBV infections of marked severity. The association of this mutation with fulminant hepatitis has been described but is at best erratic. Even if a significant association between these two observations could be demonstrated, causality did not necessarily follow as each one of them could in turn be associated with a third event, ultimately the cause of both. The observation of the postulated effect after transfection of appropriate cultured cells or infection of an experimental animal with a molecule which contains the mutation to be analysed engineered into an otherwise wild type genome will be the ultimate proof of such proposed effect on pathogenicity.

Mutations of a different kind have been described in the S gene, and more specifically within the *a* determinant region. These mutations, of which that in position 145 (Gly → Arg) is the most common have been observed mostly in child vaccinees and in HBV-infected recipients of liver transplants that have been passively immunised with large doses of HBV immunoglobulin (Carman *et al.*, 1990). Finally other cases, which are naturally occurring, tend to be patients who are serum HBsAg negative, anti-core positive, anti-sAg negative and HBV DNA positive (generally by after PCR amplification) and with normal liver biochemistry. Although the existence of such escape mutants is now well documented their clinical significance remains unclear. The apparently very slow diffusion of such mutations amongst the populations following massive immunisation campaigns and the low replication (pathogenicity) observed in naturally occurring cases appear to suggest that such variants are associated with low virulence. This could arise from the obligatory mutations which occur in the polymerase ORF (which shares the same sequence with S) if such mutations were to result in lower polymerase activity and consequently depressed viral replication. The use of *in vitro* polymerase assays which have just become available will make it possible to address such a question.

Treatment

Treatment of HBV has involved the use of interferon and nucleoside analogs. Treatment with interferon is effective in less than 40% of cases (Wong *et al.*, 1993). Response in patients with active viral replication and *e* antigenemia is accompanied by sustained undetectable viral replication, reduced inflammatory activity with arrested progression to cirrhosis, and general improvement of the quality of life. Seroconversion from *e* to anti-*e* normally results from treatment but seroconversion to anti-HBs in excess of the observed natural seroconversion rate is infrequent and often delayed. Persistence of covalently closed circular (CCC) HBV DNA molecules in the nucleus in spite of a favourable response to therapy demonstrates failure to eradicate the virus and a risk of recurrence.

The non-responder group include cases of actively replicating virus as well as low replicators (anti-*e* positive) and the molecular basis for their resistance to interferon treatment is not known. Other antivirals have met with little success, frequently because of unacceptable toxicity, with the possible exception of a nucleoside analog, the (−) enantiomer of 3′ thiacytidine, which has been shown to block reverse transcription of pregenomic RNA effectively and to have few side effects (Dienstag *et al.*, 1995).

Alternative strategies involving gene or antisense therapies are still tentative and both selective targetting of infected cells and ensuring continuous effective intracellular levels have proved major challenges.

Prevention

Immunization against HBV infection is based on a sub-viral immunogen which consists of naturally derived or genetically engineered HBsAg.

The vaccine produces antibody titres of more than 100 International Units/ ml in 90%–95% of the vaccinees but the levels drop markedly over a period of 5 years in a significant number of cases and a booster is normally required. The vaccination schedule involves three injections at times 0, 1 month and either 6 months or one year. The need for three injections and a possible booster and the cost still make the vaccine less than optimal for mass vaccination in rural areas of the world of HBV endemicity. In spite of this limitation large immunization campaigns already conducted in several areas with high HBV prevalence have produced striking results. The latest report from Taiwan, where HBV carriage rates were previously of the order of 10%, has shown a dramatic decrease in carriage rate two years after vaccination and a small but definite decline in the incidence of hepatocellular carcinoma (see below).

HBV and Hepatocarcinogenesis

Although relatively rare in European countries and in North America hepatocellular carcinoma is very common in Sub-Saharan Africa and also in the Far East where it is responsible for more than 250,000 deaths a year (Blumberg and London, 1990. The tumour frequently remains symptomatically silent until it reaches a fairly advanced stage of dissemination and hence carries a very poor prognosis.

A major factor in hepatocarcinogenesis is the association with HBV chronic infection, a clinical observation first reported in the seventies(Sherlock *et al.*, 1970) and later confirmed by epidemiological studies (Beasley and Hwang, 1984).

The role of hepatitis B virus in hepatocarcinogenesis is not clearly understood (see reviews by Rogler, 1991; Slagle *et al.*, 1992; Okuda, 1992; Buendia, 1992). Because in a large group of cases integrated HBV DNA sequences are found in the tumour cell genome many studies have focussed on the elucidation, at the molecular level, of their possible role(s) in carcinogenesis.

According to currently accepted views in viral oncogenesis, integrated viral genomic sequences have either a direct disregulatory effect on host genes controlling cell division or one which is mediated by chromosomal re-arrangements which ultimately determined the tumour phenotype. Amongst the direct effects the integrated sequences are thought to act either in *cis* or in *trans*. In the first case the physical insertion of a piece of foreign viral DNA may up-regulate (or de-repress) a host cell proto-oncogene by providing its own viral regulatory sequences (promoters or enhancers); alternatively it may be associated with the inactivation or deletion of tumour suppressor genes. When insertion of a viral regulatory sequence takes place, it either upregulates a protooncogene located in the same chromosome, or it mediates a chromosome break and translocation which may lead to the loss of critical genes or place viral regulatory sequences next to a proto-oncogene on a different chromosome. The significant role of chromosomal translocations in human cancers other than haematological malignancies has recently been recognized (Rabbitts, 1994).

As to the effect in *trans* of the integrated viral sequences this would be mediated by viral protein(s) expressed from the integrated HBV DNA causing the transactivation (and disregulation) of cellular genes involved in the control of cell division. The effect of such proteins would be to override the precise control of cell division leading to neoplastic transformation. In contrast with the viral oncogenes of acute oncogenic viruses, potential oncogenic effect of these proteins would not be exerted during acute infection during which replication from episomal genomes is known to proceed and three or more decades of chronic infection have normally to elapse before cancer develops.

HBV DNA Integration

The mechanism HBV DNA integration into the genome of the tumour liver cell is not fully understood. Most tumours contain several integrated viral sequences (Zhou *et al.*, 1988; Hino *et al.*, 1986; Tokino *et al.*, 1987; Yaginuma *et al.*, 1987) which in multicentric tumours may or may not be the same indicating mono or polyclonality. Full analysis of these sequences shows insertion through a preferential region of the viral genome, the cohesive region where the 5' ends of both strands overlap (Fig. 5), and suggests a mechanism of integration mediated by the minus strand during the process of viral replication (Hino *et al.*, 1989; Yang and Summers 1995). Although in some viral integrants none of the two ends map to the cohesive region it has been suggested that subsequent chromosomal re-arrangements may have erased the 'evidence' documenting the original insertional event. The length of the integrated viral sequences is thus variable when analysed from well established tumours and sub-genomic sequences have been found to be more common than full or nearly full genomic viral DNAs. Although sequences are normally colinear with the genomic map re-arrangements have been described which include deletions, inversions and duplications (Slagle *et al.*, 1992). Similarly the host sequences flanking the integrated viral DNA also undergo changes which include duplications at the host/viral junctions (Yaginuma *et al.*, 1985) and deletions which may be large (Rogler *et al.*, 1985). Translocations mediated by the integrated HBV DNA have also been reported (Hino *et al.*, 1986; Tokino *et al.*, 1987; Meyer *et al.*, 1990; Pineau *et al.*, 1996).

Following the model described for some retroviruses, i.e. MLV, where the provirus inserts itself next to a protooncogene which then becomes actively transcribed from the inserted viral promoter, host cell DNA sequences flanking the HBV integrated DNA were similarly analysed. In the case of HBV integration was found not to occur at a specific chromosomal site and a number of chromosomes (2, 3, 4, 5, 6, 7, 9, 11, 15, 17, 18, and X) are involved in the tumours that have been analysed so far if with preferred assignations (about 50%) to chromosomes 3, 11 and 17(6). The type of host cell DNA targetted appears also to be variable with only two instances of 'informational' DNA being involved, i.e., the genes for cyclin A and retinoic acid receptor (de The *et al.*, 1989; Wang *et al.*, 1990) and with cases of repetitive or semi-repetitive DNA (Okuda, 1992; Quade *et al.*, 1992) amongst the 'non-informational' DNA.

The possible role of HBV in inactivation or allelic losses of anti-oncogenes, or oncogene suppressor genes, has also been investigated. Results show losses in heterozigosity in a large number of tumours (Slagle *et al.*, 1992) and involving a variety of chromosomes (1, 4, 11, 13, and 17) but no association with the presence of integrated HBV DNA.

HBV Proteins in Hepatocarcinogenesis

In view of its transactivating properties (see above) the HBV X gene protein has been actively investigated for its potential role in hepatocarcinogenesis, particularly in view of its documented transactivation of heterologous promoters like those of interferon, class I major histocompatibility antigens, *c-myc* and HIV LTR. Furthermore, failure to demonstrate binding between the X protein and DNA indicates that the transactivating efect observed is mediated by an interaction with cellular transcription factors. Additional evidence suggesting the involvement of an X gene in hepatocarcinogenesis comes from the observation that X protein is able to transform stable cell lines, although not primary cultures, and to induce high incidence of liver tumours in transgenic mice expressing HB X protein in the liver when high expression vectors are used (Kim *et al.*, 1991). In a recent detailed study of tumour development in X gene transgenics preneoplastic foci were described which consisted of cells with high content of X protein and active DNA synthesis (Koike, 1995). In human tumours integrated X gene sequences are commonly present, frequently truncated at the carboxy terminus (Yaginuma, 1987; Quade *et al.*, 1992), sometimes expressed, and in some instances shown to retain transcriptional transactivating properties (Wollersheim *et al.*, 1988); Takada and Koike ,1990). So far X gene expression has not been shown to be associated with overexpression of known protooncogenes although transactivation was shown in one instance to be mediated by activation of protein kinase C (PKC) which in turn activates transcription factors AP-1 (Jun-Fos) and others (AP-2, NF-kB) and cell transformation could therefore be mediated by the PKC signal cascade. The expression of X protein has also been reported to potentiate the induction of hepatocarcinogen-mediated tumours (Slagle *et al.*, 1996).

Recently another viral transactivating sequence derived from a human tumour has been described. It consists of a preS/S (pre-surface/surface antigen) sequence truncated of its carboxyl end (Caselmann *et al.*, 1990) but its prevalence in tumours and hence general significance are not known.

The possible oncogenic role of the preS1 protein has also been suggested in lineages of transgenic mice in which high expression of the viral protein is associated with a high incidence of hepatocellular carcinomas. A progressive series of histological changes has been documented from the accumulation of non-secretable filamentous surface antigen material to cellular hyperplasia, adenomata and ultimately carcinomas. The changes appear to be initiated by the mechanical effect of the accumulated preS1 protein causing cell death which is then followed by an inflammatory and regenerative response (Chisari *et al.*, 1989). The cells which make up the hyperplastic response contain little or no viral protein and the tumours show no increase in expression of any of the tested oncogenes or of p53 (Pasquinelli *et al.*, 1992). The cells which are stuffed with preS1 material are said to be histologically similar to ground glass cells (known to contain large amounts of HBsAg) although the latter have not been shown to contain exclusively pre-S1 material or to be the precursors of neoplastic transformation. Because of the nature of the experimental model used, where all liver cells contain since birth an integrated recombinant HBV construct overexpressing PreS1, the general significance of the changes observed must await the analysis of more integrated HBV sequences from naturally occuring tumours and the detailed study of the early histological changes in natural hepatocellular carcinomas in experimental animals like the woodchuck.

An alternative model suggests that HBV is just a non-specific inducer of cell division which occurs as part of the regenerative process that follows cellular necrosis. Increased occurence of HCC (albeit with a much lower incidence) has been reported in liver diseases other than HBV infection where cirrhosis occurs with its accompanying regenerative nodules (alcoholic cirrhosis, haemochromatosis) suggesting that active cell division may be an important factor in a multi-stage oncogenic process which involves mutations and chromosome losses or re-arrangements. The recently reported high incidence of antibodies to a highly cirrhogenic RNA hepatitis virus (HCV) amongst HCC cases with no markers of HBV current infection appears to lend support to such a hypothesis. However the absence of cirrhosis in well documented cases of HCC in humans and in WHV associated hepatocellular carcinomas (Popper *et al.*, 1987) suggests that the necro-inflammatory changes associated with chronically active hepatitis rather than cirrhosis itself may be the most important factor. This would explain the absence of tumours in chronically HBV infected chimpanzees where inflammatory changes are usually limited to the portal spaces and cell necrosis is normally absent or minimal. This view also receives support from the observation of high incidence of HCC observed in the mutant LEC rat in which fulminant hepatitis ocurs four months after birth and is followed by chronic hepatitis and hepatocellular carcinoma in association with low levels of ceruloplasmin and heavy hepatic copper deposits (Ono *et al.*, 1991). The changes described in the lineages of transgenic mice where HBV preS1 protein accumulates and induces cell necrosis and regeneration already mentioned above would constitute still another observation supporting this hypothesis (Chisari *et al.*, 1989).

Hepatocarcinogenesis Associated with other Hepadnaviruses

WHV

Woodchuck hepatitis virus (WHV), another member of the hepadna family, also shows a strong association with HCC. In animals infected in the wild (one presumes predominantly through maternal transmission) the incidence is almost 100% within the first two years of captivity (Popper *et al.*, 1987). The tumour is either mono or multicentric and develops on a background of mild chronic active hepatitis but without cirrhosis.

As with HBV, WHV DNA is found covalently linked to host cell DNA in the tumour cell and in some cases has undergone major re-arrangements. Insertional activation of the *myc* family of genes (c-myc and N-myc) has been reported in about 50% of the woodchuck HCCs analysed (Moroy *et al.*, 1986; Fourel *et al.*, 1990). N-myc2 (a functional processed pseudogene) appears to be the preferred target for WHV insertion and in about 40% of tumours insertion tends to occur either upstream of the gene or in the untranslated 3' region (Fourel *et al.*, 1990; Hsu *et al.*, 1988; Wei *et al.*, 1992). Activation of N-myc2 does not appear *per se* to change the morphology of cells , it will only cause a mild proliferative response and is commonly associated with apoptosis under certain environmental conditions.Cells that survive this effect and go on to proliferate and transform appear to overcome the apoptotic effect of N-myc2 by upregulating the production of Insulin-like Growth Factor II (IGF-II) which is thought to be part of an autocrine mechanism in hepatocarcinogenesis (Ueda and Ganem, 1996).

GSHV and DHV

Integration of GSHV DNA is rare in HCCs of the ground squirrel and unrelated to increased expression of c-myc which has been reported in a significant number of cases.

    The association between duck hepatitis virus (DHV) and HCC has been reported in ducks from the Chi-Tung province in mainland China but the possible presence of hepatocarcinogenic toxins in the diet raises questions as to the cause of the tumors in these animals. Two studies have addressed this point by using HDV infected and uninfected ducklings both non-treated and treated with aflatoxin (AFB) which was administered either orally or intra-peritoneally (Cova *et al.*, 1990; Cullen *et al.*, 1990). Both studies concur that AFB is a potent hepatocarcinogen in the duck and that persistent infection with DHV has no role in tumour development in these animals under the conditions used.

    The lack of a defined X gene in the avian hepadnaviruses in contrast to the othohepadnaviruses (HBV, WHV and GSHV) may be a significant factor in explaining their different hepatocarcinogenetic potentials.

Some Unanswered Questions

In spite of a lot of progress in the last decade and a half several important aspects of the HBV biological cycle are still poorly understood.

Early events of virus infection, including cell attachment and entry, have not been characterized and slow progress in this area is undoubtedly related to the lack of a cell culture system that can be reproducibly and effectively infected with the virus. Availability of such a system would no doubt also contribute to the development of new antivirals and possibly to the development of a virion-based new generation of vaccines.

After significant recent progress in our understanding of the structure of the core particle, thanks to much improved resolution of electron cryomicroscopy, similar analysis of full particles is eagerly awaited.

As indicated in the text several molecular events during replication remain unclear. These include, amongst others, the mechanism for renewal of CCC DNA templates, the basis for the functional asymmetry of terminally repeated templates and the molecular 'mechanics' for transpositions of short oligos during replication from one end of the molecule to the other. Later events which include a conformational change in core that leads to the interaction with the envelope and topological changes in the latter also remain areas of active enquiry.

The significance of *e* antigen in the context of natural viral infection still remains a puzzle and the function of the X gene during natural infection is incompletely understood.

In spite of much information on the immune response to HBV our knowledge is still very limited as to the underlying molecular basis of persistent infection and of the immunopathogenic mechanism which is thought to be implicated.

As to the role of the virus in hepatocarcinogenesis, progress in this area has been disappointingly slow in recent years and the major questions remain unanswered. The role of integrated transactivating HBV DNA sequences requires further investigation and the woodchuck model offers promise for the study of early lesions and progression to malignancy.

Finally, in the areas of treatment and prevention, few therapeutic options are presently available. Those that are have significant limitations in relation to effectiveness and cost. By contrast mass vaccination in countries of high endemicity have shown very encouraging results in significantly reducing virus prevalence. Significant reduction in vaccine unit cost and possibly a new generation vaccine providing lasting immunity after a single dose are required if mass vaccination is to be carried out in highly endemic countries in Africa and if the infection is to be ultimately eradicated.

References

Almeida, J., Rubenstein, D. and Stott, E.J. *Lancet* 1971; **ii**: 1225–1227.

Antonucci, T. and Rutter, W.H. *J. Virol.* 1989; **63**: 579–585.

Araki, K., Miyazaki, J-I., Hino, O. *et al. PNAS* 1989; **86**: 207–211.

Arii, M., Takada, S. and Koike, K. *Oncogene* 1992; **7**: 397–403.

Babinet, C., Farza, H., Morello, D. *et al. Science* 1985; **230**: 1160–1163.

Bartenschlager, R., Kuhn, C. and Schaller, H. *Nucl. Acid. Res.* 1992; **20**: 195–202.

Benn, J. and Schneider, R.J. *PNAS* 1994; **91**: 10350–10354.

Beasley, R.P. and Hwang, L-Y. In *"Viral Hepatitis and Liver Disease"*. (eds.) Vyas, G.N., Dienstag, J.L. and Hoofnagle, J.H. Grune and Stratton Inc., Orlando, Fla. 1984 p. 209–224.

Blumberg, B.S., Alter, H.J. and Visnich, S. *JAMA* 1965; 541–546.

Blumberg, B.S. and London, W.T. *Cancer* 1990; **50**: 2657–2665.

Bottcher, B., Wynne, S.A. and Crowther, R.A. *Nature* 1997; **386**: 88–91.

Brunetto, M.R., Stemmler, M., Schodel, F. *et al.* Italian *J. Gastroent.* 1989; **21**: 151–154.

Bruss, V. and Ganem, D. *PNAS* 1991; **88**: 3813–3820.

Buckold, V.E., Xu, Z., Chen, M. *et al. J. Virol.* 1996; **70**: 5845–5851.

Buendia, M.A. *Adv. Cancer Res.* 1992; **59**: 167–226.

Carman, W.F., Jacyna, M.R., Hadzyiannis, S. *et al. Lancet* 1989; **ii**: 588–591.

Carman, W.F., Zanetti, A.R., Karayiannis, P. *et al. Lancet* 1990; **336**: 325–329.

Caselmann, W.H., Meyer, M., Kekule, A. *et al. PNAS* 1990; **87**: 2970–2974.

Cattaneo, R., Will, H., Hernandez, N. and Schalle, H. *Nature* 1983; **305**: 336–338.

Chang, L-J., Pryciak, P., Ganem, D. and Varmus, H.V. *Nature* 1989; **337**: 364–368.

Chang, H.K., Wang, B.Y., Yuh, C.H. *et al. Mol. and Cell Biol.* 1989; **9**: 5189–5197.

Charnay, P., Pourcel, C., Louise, A. *et al. PNAS* 1979; **76**: 2222–2226.

Cheong, J-H., Yi, M-K., Liu,Y. and Murakami, S. *EMBO J.* 1995; **14**: 143–150.

Chen, I-H., Huang, C-J. and Ting, L.P. *J. Virol.* 1995; **69**: 3647–3657.

Cherrington, J., Russnak, R. and Ganem, D. *J. Virol.* 1992; **66**: 7589–7596.

Chisaka, O., Araki, K., Ociya, T. *et al. Gene* 1987; **60**: 183–189.

Chisari, F.V., Klopchin, K., Moriyama, T. *et al. Cell* 1989; **59**: 1145–1156.

Choo, K-B., Liew, L-N., Ching, K-Y. *et al. Virology* 1991; **182**:785–792.

Chu, C-M. and Liao, Y-F. *Gastroenterology* 1987; **92**: 220–225.

Conway, J.F., Cheng, N., Zlotnick, A. *et al. Nature* 1997; **386**: 91–94.

Cova, L., Wild, C.P., Mehrotra, R. *et al. Cancer Res.* 1990; **50**: 2156–2163.

Crowther, R.A., Kiselev, N.A., Bottcher, B. *et al. Cell* 1994; **77**: 943–960.

Cullen, J.M., Marion, P.L., Sherman, G. *et al. Cancer Res.* 1990; **50**: 4072–4080.

Dane, D.S., Cameron, C.H. and Briggs, M. *The Lancet* 1970; **i**: 695–698.

Delius, H., Gough, N.M., Cameron,C.H. and Murray, K. *J. Virol.* 1983; **47**: 337–343.

De The, H., Marcho, A. and Tiollais, P. *Nature* 1989; **330**: 660–670.

Dienstag, J.L., Perrillo, R.P., Schiff, E.R. *et al. N. Engl. J. Med.* 1995; **333**: 1657–1661.

Doria, M., Klein, N., Lucito, R. and Schneider, R.J. *EMBO J.* 1995; **14**: 4747–4757.

Eble, B.E., Mac Rae, D.R., Dubois, M.F. *et al. PNAS* 1980; **77**: 4549–4553.

Eckardt, S.G., Milich, D.R. and McLachlan, A. *J. Virol.* 1991; **65**: 575–582.

Fetelson, M., Zhu, M., Duan, L.X. and London, W.T. *Oncogene* 1993; **8**: 1109–1117.

Fouillot, N., Tlouzeau, S., Rossignol, J. and Jean-Jean, O. *J. Virol.* 1993; **67**: 4486–4495.

Fourel, G., Trepo, C., Bougueleret, B. *et al. Nature* 1990; **347**: 294–298.

Fowler, M.F., Monjardino, J., Tsiquaye, K.N. *et al. J. Med. Virol.* 1984; **13**: 83–91.

Garcia, A., Ostapchuk, P. and Hearing, P. *J. Virol.* 1993; **67**: 3940–3950.

Guerrero, E., Gavilanes, F. and Peterson, D.L. In *Viral Hepatitis and Liver Disease*, (ed.) Zuckerman, A.J. Alan Liss, New York 1988 pp. 606–613.

Guo, W., Chen, M., Yeu, T.S.B. and Ou, J-H. *Mol. Cell Biol.* 1993; **13**: 1443–1448.

Heermann, K.H., Goldmann, U., Schwartz, W. *et al. J. Virol.* 1984; **52**: 396–402.

Hertogs, K., Leenders, W.P.J., Depla, E. *et al.Virology* 1993; **197**: 549–547.

Hino, O., Shows, T. and Rogler, C.E. *PNAS* 1986; **83**: 8338–8342.

Hino, O., Ohtake, K. and Rogler, C.E. *J. Virol.* 1989; **63**: 2638–2643.

Hsu, T.Y., Monroy, T., Etiemble, J. *et al. Cell* 1988; **55**: 627–635.

Hu, K., Vierling, J. and Siddiqui, A. *PNAS* 1990; **87**: 7140–7144.

Hu, K. and Siddiqui, A. *Virology* 1991; **181**: 721–726.

Huang, J., Kwong, J., Sun, E.C-Y. and Liang, T.J. *J. Virol.* 1996; **70**: 5582–5591.

Jean-Jean, O., Weimer, T., de Recondo, A.M. *et al. J. Virol.* 1989; **63**: 5451–5454.

Katayama, K., Hayashi, N., Sasaki, Y. *et al. Gastroenterology* 1989; **97**: 990–998.

Khudyakov, Y.E. and Makhov, A.M. *FEBS* 1989; **243**: 115–118.

Klingmuller, U. and Schaller, H. *J. Virol.* 1993; **67**: 7414–7422.

Kim, C-M., Koike, K., Saito, I. *et al. Nature* 1991; **351**: 317–320.

Kock, J., Borst, E-M. and Schlicht, H-J. *J. Virol.* 1996; **1970**: 5827–5831.

Koike, K. *Intervirology* 1995; **38**: 134–142.

Kuroki, K., Cheung, R., Marion, P. and Ganem, D. *J. Virol.* 1994; **68**: 2091–2096.

Landers, T.A., Greenberg, H.B. and Robinson, W.S. *J. Virol.* 1977; **23**: 368–376.

Liao, W. and Ou, J-H. *J. Virol.* 1995; **69**: 1025–1029.

Loeb, D.D., Hirsch, R.C. and Ganem, D. *EMBO J.* 1991; **10**: 3533–3540.

Marion, P.L., Oshiro, L.S., Regnery, D.C. *et al. PNAS* 1980; **77**: 2941–2945.

Mason, W.S., Seal, G. and Summers, J. *J. Virol.* 1980; **36**: 829–836.

Medi, H., Kaplan, M., Anlar, F. *et al. J. Virol.* 1994; **68**: 2415–2424.

Meyer, M., Wiedorn, K.H. and Hofschneider. *Hepatology* 1992; **15**: 665–671.

Monjardino, J., Fowler, M.F., Montano, L. *et al. J. Med. Virol.* 1982; **9**: 189–199.

Moriyama, T., Guilhot, S., Klopchin, K. *et al. Science* 1990; **248**: 361–364.

Moroy, T., Marchio, A., Etiemble, J. *et al. Nature* 1986; **324**: 276–279.

Nakamura, I. and Koike, K. *Virology* 1992; **191**: 533–540.

Neurath, A.R., Kent, S.B.H., Strick, N. and Parker, K. *Cell* 1986; **46**: 429–436.

Norder, H., Carouce, A-M. and Magnius, L.O. *J. Gen. Virol.* 1992; **73**: 3141–3145.

Offensperger, W-B., Offensperger, S., Walter, E. *et al. Virology* 1991; **183**: 415–418.

Okamoto, H., Tsuda, F., Sakugawa, H. *et al. J. Gen. Virol.* 1988; **69**: 2575–2583.

Okamoto, H., Tsuda, Y., Akahane, Y. *et al. J. Virol.* 1994; **68**: 8102–8110.

Okuda, K. *Hepatology* 1992; **15**: 948–963.

Ono, T., Abe, S. and Yoshida, M.C. *Jpn. J. Cancer Res.* 1991; **82**: 486–489.

Ostapchuk, P., Hearing, P. and Ganem, D. *EMBO J.* 1994; **13**: 1048–1057.

Pasquinelli, C.C., Bharani, K., and Chisari, F.V. *Cancer Res.* 1992; **52**: 2823–2829.

Pontisso, P., Petit, M.A., Vankowski, M. and Peeples, M.J. *Virol.* 1989; **63**: 1981–1988.

Prange, R. and Streeck, R.E. *Embo J.* 1995; **14**: 247–256.

Qadri, I., Maquire, H.F. and Siddiqui, A. *PNAS* 1995; **92**: 1003–1007.

Rabbitts, T.H. *Nature* 1994; **372**: 143–149.

Radziwill, G., Zentgraft, H., Schaller, H. and Bosch, V. *Virology* 1988; **163**: 123–132.

Radziwill, G., Tucker, W. and Schaller, H. *J. Virol.* 1990; **64**: 613–620

Rigg, R.J. and Schalle, H. *J. Virol.* 1992; 66: 2829–2836.

Robinson, W.S. and Greenman, R.L. *J. Virol.* 1974a; **13**: 1231–1236.

Robinson, W.S., Clayton, D.A. and Greenman, R.L. *J.Virol.* 1974b; **14**: 384–391.

Rogler, C.E., Sherman, M. and Su, C.Y. *et al. Science* 1985; **230**: 319–322.

Rogler, C.E. *Current Topics in Microbiology and Immunology* 1991; **168**: 103–140.

Roossnck, M.J. and Siddiqui, A. *J. Virol* 1987; **61**: 955–961.

Roychoudhury, S. and Shih, C. *J.Virol.* 1990; **64**: 1063–10069

Schlicht, H-J., Baterschlager, R. and Schaller, H. *J. Virol.* 1989a; **63**: 2995–3000.

Schlicht, H., Radziwill, G. and Schaller, H. *Cell* 1989b; **56**: 85–92.

Sells, M.A., Chen, M-L. and Acs, G. *PNAS* 1987; **84**: 1005–1009.

Seto, E., Yen, T.S.B., Peterlin, B.M. and Ou, J-H. *PNAS* 1988; **85**: 8286–8296.

Sherlock, S., Fox, R.A., Niazi, S.P. *et al. Lancet* 1970; **i**: 1243–1247.

Shin, H.J. and Rho, H.M. *J. Biol. Chem.* 1995; **270**: 11047–11050.

Slagle, B.L., Lee, T-H. and Butel, J. *Prog. Med. Virol.* 1992; **39**: 167–203.

Slagle, B.L., Lee, T.H., Medina, D. *et al. Mol. Carcinog.* 1996; **15**: 261–269.

Spandau, D.F. and Lee, C.H. *J. Virol.* 1988; **62**: 427–434.

Sprengel, R., Kaleta, E.F. and Will, H. *J. Virol.* 1988; **62**: 3832–3839.

Standring, D.N., Rutter, W.J., Varmus, H.E. and Ganem, D. *J. Virol.* 1984; **50**: 563–571.

Summers, J. and Mason, W. *Cell* 1982; **29**: 403–415 .

Summers, J., Smolec, J.M. and Snyder, R. *PNAS* 1978; **75**: 4533–4537.

Summers, J., O'Connell, A. and Millman, I. *PNAS* 1975; **72**: 4597–4601.

Takada, S. and Koike, K. *PNAS* 1990; **87**: 5628–5632.

Takahashi, K., Aoyama, K., Ohno, N. *et al. J. Gen. Virol.* 1955; **76**: 3159–3164.

Takada, S. and Koike, K. *PNAS* 1990; **87**: 5628–5632.

Tavis, J., Perri, S. and Ganem, D. *J. Virol.* 1994; **68**: 3568–3594.

Toh, H., Hayashida, H. and Miyata, T. *Nature* 1983; **305**: 827–829

Tokino, T., Fukushige, S., Nakamura, T. *et al. J. Virol.* 1987; **61**: 3848–3854.

Trujillo, M.A., Letovsky, J. and Maguire, H.F. *PNAS* 1991; **88**: 3797–3801

Tutleman, J., Pourcel, C. and Summers, J. *Cell* 1986; **47**: 451–460.

Twu, J.S. and Schloener, R.H. *J. Virol.* 1987; **61**: 3448– 3453.

Ueda, K. and Ganem, D. *J. Virol.* 1996; **70**: 1375–1383.

Wang, J., Chenivesse, X., Henglein, B. and Brechot,C. *Nature* 1990; **343**: 555–557.

Wang, Y., Chen, P., Wu, X. *et al. J. Virol.* 1990; **64**: 3977–3981.

Wang, G-H. and Seeger, C. *J. Virol.* 1993; **67**: 6507–6512.

Wey, W., Fourel, G., Ponzetto, A. *et al. J. Virol.* 1992; **66**: 5265–5276.

Wey, Y., Tavis, J.E. and Ganem, D. *J. Virol.* 1996; **70**: 6455–6459.

Wollersheim, M., Debelka, U. and Hofschneider, P.H. *Oncogene* 1988; **3**: 545–552.

Wong, D.K.H., Cheung, A.M., Naylor, C.D. *et al. Ann. Intern. Med.* 1993; **119**: 312– 323.

Yaginuma, K., Kobayashi, M., Yoshida, E. and Koike, K. *PNAS* 1985; **82**: 4458–4462.

Yaginuma, K., Kobayashi, H., Kobayashi, M. *et al. J. Virol.* 1987; **61**: 1808–1813.

Yaginuma, K., Nakamura, I., Takada, S. and Koike, K. *J. Virol.* 1993; **67**: 2559–2565.

Yang, W. and Summers, J. *J. Virol.* **69**: 4029–4036.

Yang, W., Mason, W.S. and Summers, J. *J. Virol.* 1996; **70**: 4567–4575.

Yee, J. *Science* 1989; **246**: 658–670.

Yeh, C-T., Liaw, Y-F. and Ou, J-H. *J. Virol.* 1990; **64**: 6141–6147.

Yen, T.S.B. *Semin. Virol.* 1993; 4133–4142.

Yu, X. and Mertz, J. *J. Virol.* 1996; **70**: 8719–8726.

Yu, M. and Summers, J. *J. Virol.* 1994; **68**: 2965–2969.

Zhou, Y.Z., Slagle, B.L., Donehower, L.A. *et al. J. Virol.* 1988; **62**: 4224–4231.

PART 3

DELTA HEPATITIS VIRUS

History

Hepatitis delta virus was first detected in 1977 as a new hepatocyte nuclear antigen in patients infected with hepatitis B virus and was shown to be frequently associated with acute fulminant or rapidly progressing chronic disease (Rizzetto *et al.*, 1977).

The antigen (HDAg) was first shown to be un-related to HBV and later to be associated with a new agent, an RNA-containing virus enveloped in HBV surface antigen (Rizzetto *et al.*, 1980). Detection of HDAg in liver and assays for detection of serum antibodies and viral antigen made possible the diagnosis of HDV infection and the characterization of its natural history. Analysis of the virus at the molecular level, including the cloning of its genome, have since made important contributions to our understanding of virus replication, assembly, virus detection and pathogenesis.

The Virus

Classification

HDV has a circular single-stranded RNA genome of negative polarity. The virus has yet to be classified. The structure of its genome and the mechanism of its replication show striking similarities with some plant pathogens (viroids, virusoids and satellite RNAs). It is however clearly distinct from viroids in containing a larger genome which encodes one protein, delta antigen (HDAg), and in possessing an envelope (Riesner and Gross, 1985). The agent is also quite different from virusoids and satellite RNAs, the former being similar in size and structure to viroids but encapsidated together with single-stranded plant RNA viruses and the latter enveloped in helper virus protein coats (as HDV) but incapable of autonomous replication.

The agent is the smallest human pathogen known and the only example of a human virus containing a circular RNA genome.

The Virion

HDV is a small RNA virus consisting of spherical particles about 36 nm in diameter and with a buoyant density of 1.25 gm/cm^3 (Bonino *et al.*, 1984). Particles of similar size and density were derived from cells transiently transfected *in vitro* with HDV cDNA and HBsAg (Wang *et al.*, 1990) but showed a larger diameter (40 nm) and a significantly lower buoyant density of 1.19 gm/cm^3 when HDV DNA was co-transfected with WHsAg (Ryu, W-S, *et al.*, 1992). The outer coat of the particle is made of HBsAg which consists almost exclusively of the major HBsAg polypeptide and less than 5% of preS1 and preS2 (Bonino *et al.*,1986). The coat surrounds the genome and two forms of delta antigen: small (S-HDAg) and large (L-HDAg) with molecular weights of 24,000 and 27,000 daltons, respectively. The number of antigen molecules estimated at 70 per virion (Ryu *et al.*, 1993) and the variable stoichiometry of both forms of antigen (Ryu *et al.*, 1992) point against an ordered internal nucleocapsid structure of icosahedral symmetry which EM studies of serum-derived particles have also failed to confirm. The possibility that two particle populations may exist,each containing one type of antigen, appears unlikely since it has now been demonstrated that the large form of the antigen is essential for packaging virions (see below).

The Genome

HDV genome is a circular single-stranded RNA molecule of negative polarity, about 1700 nucleotides long, of high (60%) G+C content and with up to 70% intra-molecular base-pairing. The latter accounts for its rod-shape when seen by EM under non-denaturing conditions (Kos *et al.*, 1986) and its electrophoretic mobility. These properties are shared with the RNAs of some plant pathogens, viroids virusoids and satellite RNAs (Riesner and Gross, 1985). Viroid consensus sequences GAAAC and GAUUUU as well as a 21 nucleotide-long palindrome are also found in the HDV genome.

The genome appears to contain only one common functional Open Reading Frame (ORF) in the antigenomic strand which maps between nucleotides 1600 and 959/1015 of Wang's sequence (Wang *et al.*, 1987). In addition the genome also contains a viroid-like domain mapping between nucleotides 613 and 980 (Branch *et al.*, 1989). Both domains are shown in diagrammatic form in Fig. 13.

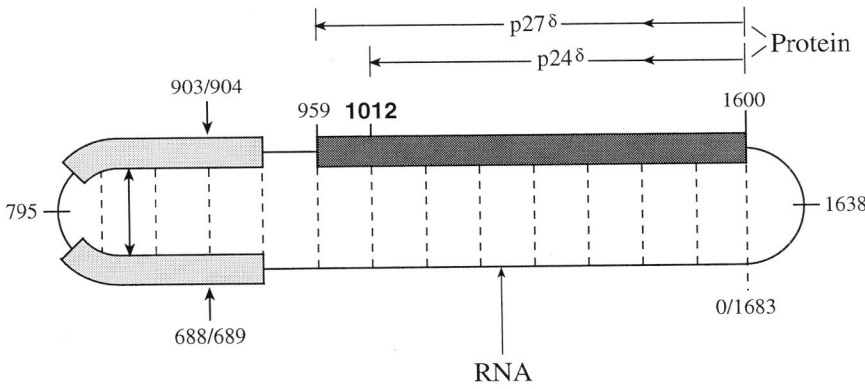

Fig. 13 HDV genome. ▨ : viroid-like region; ▦ : Open Reading Frame. p27d and p24d: two forms of HDAg (adapted from Monjardino and Lai, 1995).

Nine distinct full length and several partial sequences of the HDV genome from around the world reported so far appear to segregate into three major genotypes: type I or Italian which is the type most commonly found in Western Europe, North America, Middle East, South Pacific and Asia (Wang *et al.*, 1986; Makino *et al.*,1987; Kuo *et al.*,1988; Saldanha *et al.*, 1990; Kos *et al.*, 1991; Lee *et al.*, 1992; Chao *et al.*, 1991); type II, which was found in Japan and Taiwan (Imazeki *et al.*, 1990; Wu *et al.*, 1995); and type III which has been described in South America (Casey *et al.*, 1993). Genetic variation is highest between group III and groups I and II (60–65%), and is around 25% between the latter. A further subdivision of group I into two subtypes IA (possibly an Asian subtype) and IB has been proposed (Casey *et al.*, 1993). Sequence relatedness between isolates is shown in Fig. 14.

Sequence diversity appears to occur predominantly within the region between nucleotides 1 and 615 although frequent nucleotide changes have also been described within the ORF (Chao *et al.*, 1990). High degree of sequence conservation is seen in some regions of the genome like those between nucleotides 682–738 and 859–907 which include both genomic and anti-genomic cleavage domains, and the middle portion of the coding region. Only three regions of more than 20 conserved nucleotides have been identified amongst all the isolates that has so far been sequenced.

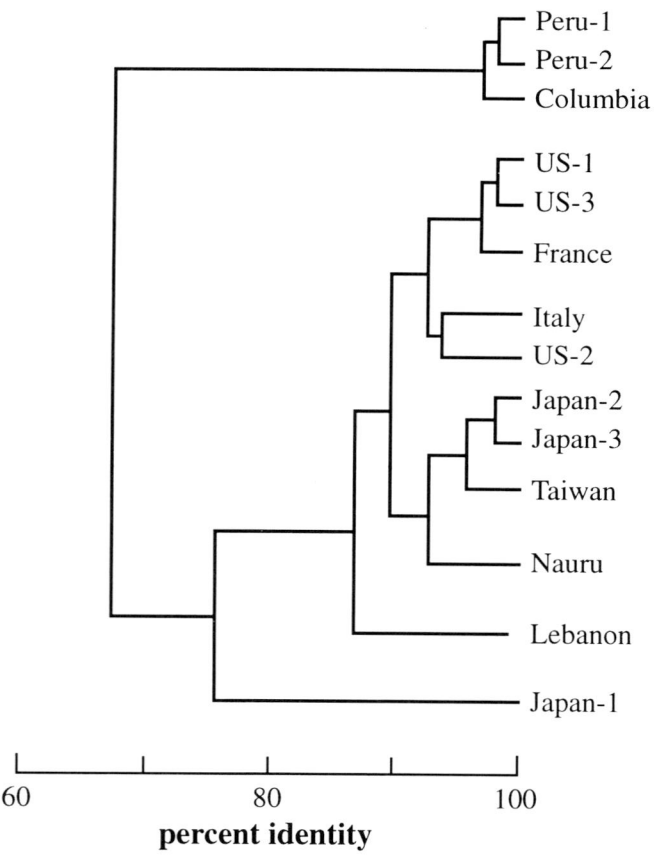

Fig. 14   Genetic relatedness amongst HDV isolates (adapted from Casey *et al.*, 1993).

Genomic sequence variation in a single patient (microheterogeneity) and throughout the course of the disease has been reported (Wang *et al.*, 1986; Lee *et al.*, 1992; Chao *et al.*, 1991). By contrast, in two full sequences originating from a single chimpanzee isolated and serially passaged, either in the chimpanzee or both in the chimpanzee and the woodchuck, very few nucleotide changes were recorded with 99.8% and 99.5% homologies, respectively (Kos *et al.*, 1991; Kuo *et al.*, 1988). A similar finding has been reported after serial passaging in woodchuck of virus obtained from cells transfected with cloned HDVcDNA (Netter *et al.*, 1995). Preferential occurrence of non-synonymous (amino acid altering) mutations over synonymous (amino acid non-altering) mutations is difficult to explain by the neutral theory of molecular evolution (Kimura, 1968) and appears to imply some selection for amino acid changes but also to reflect strong preference for C and G nucleotides at third codon positions (Krushkal and Li, 1995).

HDV Antigen

The RNA genome contains one Open Reading Frame (ORF) which encodes the HDV antigen (HDAg), a viral marker which is found in two forms (24 kD or short and 27 kD or long) both in serum and in infected liver. The two forms of HDAg are amino co-terminal and differ at the carboxyl terminus where as a result of a mutation at position 1012 ($A \rightarrow G$), a termination codon changes to tryptophan and a further 19 amino acids ($195 \rightarrow 214$) can be read through to the next termination codon. The finding of two genomes, each encoding one of the two forms of the antigen has excluded the presence of a tRNA suppression mutation; and the observed mutation of the genome encoding the small HDAg to the form encoding the large antigen both after *in vitro* and *in vivo* transfection has conclusively implicated an RNA editing mechanism. After contradictory claims it is now thought that the antigenomic strand is edited first and that an adenosine at position 1012 is deaminated to inosine. This in turn hydrogen-bonds to cytidine in the genomic strand when it is transcribed and the latter will ultimately basepair to guanosine when new antigenomic strand is synthesised (see Replication below). Recent reports appear to implicate adenosine deaminase (DRAD), or another member of a now emerging family of related proteins (Melcher *et al.*, 1996) as the HDV 'editase' enzyme. This is based on its requirement for double-stranded regions flanking the editing site of the RNA substrate and on evidence that DRADA can edit HDV RNA *in vitro* (Polson *et al.*, 1996). The requirement for double-stranded regions flanking the editing site has been documented by analysing the effect on editing of mutated bases in both flanks using an *in vitro* editing assay (Casey *et al.*, 1992) and by demonstrating *in vivo* that a mutation at position 1014 is accompanied by a mutation in the nucleotide that lies opposite in the rod conformation, i.e., nucleotide 578, which re-establishes the Watson-Crick base-pair (Wang *et al.*, 1995).

The two forms of the antigen appear to have distinct properties, the smaller molecule being required for viral replication whereas the larger molecule is inhibitory of replication but required for packaging and subsequent export of progeny particles. Functional regions mapped on the HDAg molecule include an RNA interacting domain, a protein-protein interacting domain, and a nuclear targetting domain (Fig. 15).

The RNA-binding region maps between amino acids 89–163 (Lin *et al.*, 1990), an adjacent region which maps between amino acids 67–88 contains a nuclear localizing region (NLR) of the type K(K,R)X(K,R) and the amino-terminal region, aas 1–57, contains a leucine-zipper or coiled-coil structured region which has been shown to be involved in dimerization (Xia *et al.*, 1992) (Fig. 15).

One domain is specific to the L-form of HDAg, i.e. the additional carboxy-terminal 19 amino acids. Within this sequence the cysteine residue at position 211 has been shown to be isoprenylated, and its replacement with serine stops

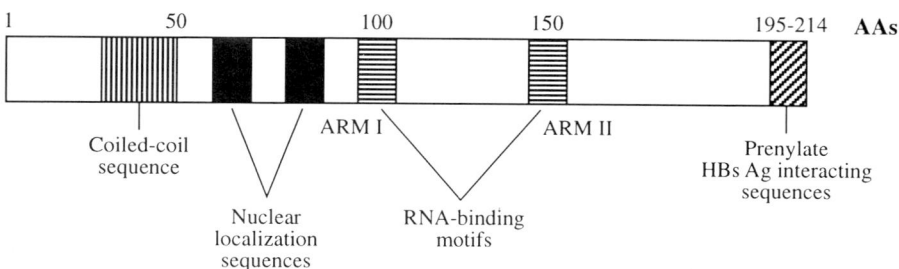

Fig. 15   Different functional domains of HDAg. AAs: amino acids.

virus packaging (Glenn *et al.*, 1992). The isoprenylation motif (CXXX, where X can be any amino acid) and the trans-inhibition of replication, which is a specific property of the large form of the antigen, have been shown to map to distinct domains since mutations that fail to affect packaging can still abolish inhibition of replication (Glenn and White *et al.*, 1991) and others that can affect packaging have no effect on viral replication (Chen *et al.*, 1996). In fact the trans-inhibitory effect does not appear to depend on the presence of the 19 additional carboxy-terminal amino acids of the L-form since fusion proteins which do not include these amino acids as part of the HDAg moiety inhibit replication as efficiently as the wild type protein (Xia and Lai, 1992). Residues between 1–68 ,on the other hand, appear to be both necessary and sufficient to confer this phenotype (Lazinsky and Taylor, 1993). Hence both this region and that implicated in the stimulation of viral replication by the small form of the antigen, are common to both forms and a specific effect associated only with one, small or large, but not the other would require conformational changes, presently undefined, whereby domains are differentially exposed.

As first described in the original report by Rizzetto *et al.*, in 1977 cellular distribution of the antigen is almost exclusively nuclear. Three patterns have since been described which include accumulation limited to the nucleoplasm, to the nucleoli, or seen in both. Recent evidence based on confocal immunofluorescence microscopy has shown that in the presence of HDV replication HDAg accumulates preferentially in the nucleoplasm and clearly disproved previous claims of differential distribution of the two forms of antigen (Cunha *et al.*, submitted 1998). Antigen accumulations in the form of minute speckles have also been described in the nucleus and their significance is unclear (Fig. 16). A recent report suggests that they may co-localize with speckles similar in size and distribution of a splicing factor SC35 (Bichko and Taylor, 1996). In another study, using confocal microscopy, HDAg speckles were shown to be in close proximity to, but not to co-localize with SC35 speckles or with any of a number of transcription and mRNA processing markers (Cunha *et al. in press*).

In view of the strong natural immunogenicity of HDAg, epitope mapping has been carried out by several groups either using synthetic peptides or variously truncated recombinant antigen molecules. Initial reports of a variety

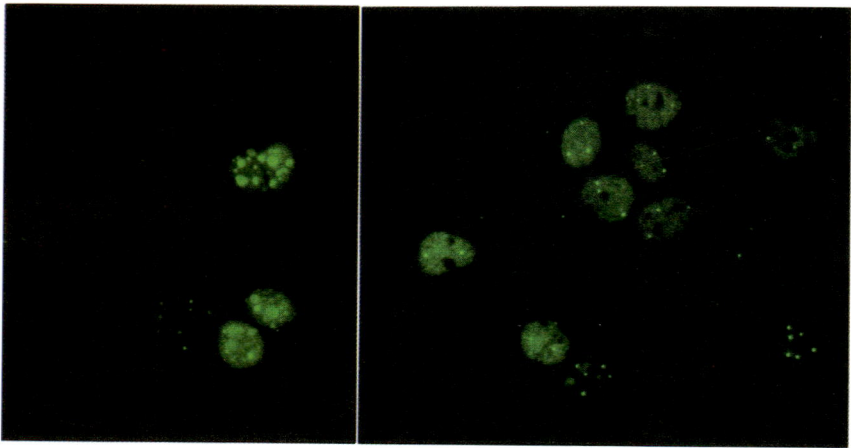

Fig. 16 HDAg expression in stably transfected cells. Immuno fluorescence staining of HDAg shows nucleoplasmic and nucleolar patterns of expression. Additional distinct punctate accumulations of antigen can also be seen (see text). (Cheng and Monjardino, unpublished).

of epitopes scattered throughout the molecule have not been confirmed and only a major common epitope located near the amino-terminus (amino acids 13–76) appears to have been unambiguously identified (Kuo *et al.*, 1988; Saldanha *et al.*, 1990; Bergman *et al.*, 1989).

Using the yeast two-hybrid assay Brazas and Ganem have recently identified a cellular protein that interacts with HDV antigen. (Brazas and Ganem, 1996). The protein designated DIPA (delta-interacting protein A) is of similar size to HDAg (202 amino acids vs. 195 or 214) and shows 24% amino acid homology, which increases to 56% if conservative amino acid changes are included. DIPA has two leucine heptads with the potential to form coiled-coil protein interaction domains, and one potential nucleic acid interacting domain (as opposed to two in HDV). Its sequence is compatible with it being a nucleic acid binding protein, even a transcription factor, although its function in non-infected cells is presently unknown. DIPA mRNA has been detected in both HepG2 hepatoma cells and in a variety of human tissues including the liver. Interaction between DIPA and HDAg has been confirmed in cultured cells expressing both proteins and requires the coiled-coil domains of both proteins. In addition DIPA has been shown to be a potent inhibitor of HDV replication and HDAg relieves such inhibition by interacting with DIPA.

Replication of HDV Genome

Replication of HDV RNA takes place in the nucleus of the infected hepatocyte and is thought to occur by the rolling circle mechanism described for other viruses and also for viroids (Chen *et al.*, 1986; Dourakis *et al.*, 1991) whereby an antigenomic strand is first copied and ligated to generate a replicative form from which genomic RNAs are ultimately synthesized (Fig. 17). As new nucleotides are added on to the 3′-OH the 5′ end of the nascent strand 'rolls off' before being cut to monomer size by the intrinsic self-cleaving (ribozyme) activity of the RNA and sealed by its self-ligating activity. A double rolling circle mechanism has been postulated to accommodate the finding of circular dimers and trimers of HDV RNA of both polarities in infected chimpanzee and woodchuck livers (Chen *et al.*, 1986; Dourakis, 1991). Monomer-size HDV RNA molecules are present in the liver at about 300,000 copies per cell and

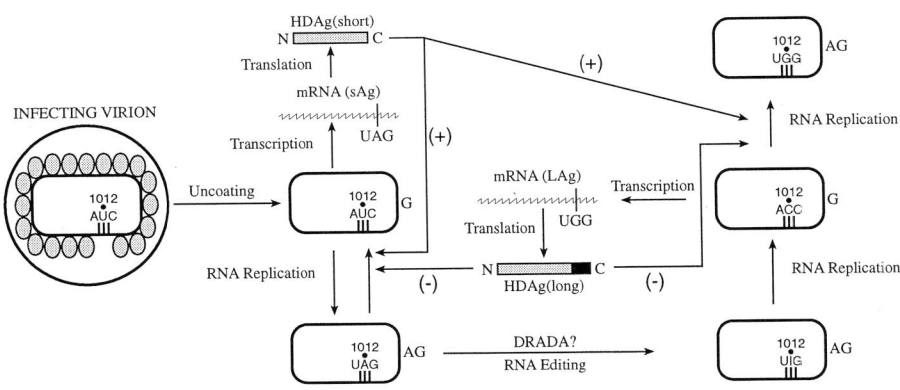

Fig. 17 HDV replication. ◯: delta antigen in full particle; ▬▬▬ : genomic/antigenomic RNA; ⌇⌇⌇⌇⌇ : mRNA; ▭▭ : HDAg; (+): upregulation; (−): downregulation; DRADA: ds RNA adenosine deaminase .

about 10^9–10^{11} particles per ml of serum; and all forms of HDV RNA are absent from all other tissues so far analysed demonstrating the strict hepatotropism of the virus (Negro *et al.*, 1989; Dourakis *et al.*, 1991). The significance of the multimeric RNA molecules is presently unclear since they can either be natural replicative intermediates prior to processing or 'junk' RNA which cannot be processed further, an explanation which appears to be contradicted by their discrete multimeric size.

Both the self-cleavage (ribozyme) reaction and the self-ligation have been shown to take place in the absence of protein and to be regulated by the concentration of magnesium ions. The precise sites for autocatalytic cleavage have been precisely mapped to the phosphodiester bonds between nucleotides 900/901 for the genomic strand (Kuo *et al.*, 1988b; Wu and Lai, 1989) and nucleotides 685/686 for the antigenomic strand (Kuo *et al.*, 1988b; Sharmeen *et al.*, 1988) (Fig. 13). Minimum length of the flanking sequences, both 5' and 3', required for cleavage have been determined as 1 and 85 nucleotides respectively, and essential structural elements have been identified. These include two hairpin loops and possibly some stabilizing tertiary interaction for increased cleavage by the HDV ribozyme which is structurally different from the 'hammerhead' and 'hairpin' types previously described. The hydrolytic reaction results in a free 5' hydroxyl and 2'–3' monophosphate ends and is followed by self-ligation of the newly-generated monomeric RNAs which will only take place if the free ends of the molecule are held in close proximity by intramolecular base-pairing.

Apart from the genomic and multi-genomic sized HDV RNAs, a 0.8 kb transcript has been described which is initiated from a promoter-like region close to what is conventionally described as the 'top' end of the rod (Figs. 13 and 17) and terminated at the unique termination signal AAUAAA which is located about 60 nucleotides downstream from the termination codon. Since both this transcript and the genomic size molecules have been shown to share a 5' end, a mechanism must operate in order to extend transcription beyond the termination signal when full length and multimeric molecules are synthesised. Because HDAg is required for replication and will control replication when supplied in *trans* its role as an antiterminator has been postulated in the absence of *in vitro* transcription data.

In spite of the significant expression of HDAg in liver cells many attempts to detect the 0.8 kb HDAg mRNA in HDV infected tissues have proved unsuccessful and in the few successful reports recombinant HDV subgenomic constructs were almost invariably used. Recently a proposal has been put forward to suggest that the RNA template for the bulk of HDAg synthesis is, in fact, full length antigenomic HDV RNA and that only minute amounts of 0.8 kb species may be required for the synthesis of HDAg as antiterminator. Once this is achieved further synthesis would be prevented by the presence of HDAg 'sitting' on the termination site and both its renewal and the bulk synthesis of HDAg for the making of progeny particles would be directed by full length antigenomic HDV RNA. Such a prediction should be easily testable *in vitro* although circular RNA molecules would not be anticipated to work effectively as mRNAs in the absence of an Internal Ribosome Entry Site (IRES).

Another unanswered question relates to the nature of the enzyme involved in the replication of HDV. Present evidence suggests that it is the host enzyme RNA polymerase II, based on the inhibition of HDV RNA replication observed in transfected cells and nuclear extracts in the presence of α-amanitin and of polymerase II-specific antibody (MacNaughton *et al.*, 1990; Fu and Taylor, 1993). Although normally transcribing a double-stranded DNA template, it has been suggested that its ability to use HDVRNA (and also viroid RNA) may depend on the double-stranded, almost DNA-like, structure of the viral genome. Again the use of the purified enzyme *in vitro*, possibily in the presence of liver transcription factors, should allow this hypothesis to be addressed experimentally.

The presence of HDVRNA in a double-stranded rod-like configuration inside the cell might have suggested a significant role for interferon in controlling HDV infection as *ds* RNA is known to be a powerful interferon inducer. However, the very fact that active virus replication can occur in infected cells indicates that it either fails to induce an effective interferon response or that it evades any such response. The latter is supported by the observation that both IFN-alpha and IFN-gamma administered to stable HDV cDNA transfected cells fail to inhibit viral replication in cells with otherwise intact interferon-responsive pathways, including the dsRNA activation of protein kinase PKR (McNair *et al.*, 1994). This has been confirmed in a recent study which reports HDV RNA activation of PKR but demonstrates its failure to inhibit protein synthesis (Robertson *et al.*, 1996). Although the reason why HDV replication in HDV-transfected cells is refractory to interferon is not known this may well mimic the poor response to interferon therapy in chronic HDV infection.

Virion Assembly and Release

As reviewed before, when discussing the properties of HDVAg, once the genomic RNA has been replicated it will bind S-HDAg and this in turn will bind L-HDAg with variable stoichiometry. Why antigenomic HDVRNA, which can also bind both forms of HDAg, is not assembled, is presently not clear. The nucleoprotein complex eventually exits the nucleus by a process which may be facilitated by a LHDAg determinant displayed on the complex.

The subsequent assembly of the virions which involves an interaction between HDV RNA-protein complex and HBsAg is still poorly understood. This requires the coming together of components that lie in different compartments of the cell, S- and L-HDAg in the nucleus and HBsAg in the cytoplasm. In experiments where both HBV envelope and HDAg genes are transfected and particles are monitored in the culture medium, it has been shown that virion assembly occurs with the small HBsAg polypeptide alone, that the middle HBsAg sequence (preS2) is dispensable but that only the addition of the large HBsAg (preS1) will produce infectious particles (Sureau, 1992). As with HBV this suggests that there is a preS1 (L protein) hepatocyte-specific receptor which mediates cell entry but differences in receptor structure will have to be postulated to explain the infectivity shown by HDV and non-infectivity by HBV of the same primate primary hepatocyte cultures (Sureau *et al.*, 1991). Transfection experiments have demonstrated that this last stage of the virion assembly process, i.e., the inclusion of HDAg in an HBV envelope, does not appear to require the HDVRNA-HDAg complex since particles can be assembled in the absence of HBV RNA. Whether such particles are formed during natural infection is not known although reports that HDAg was positive in 57% of HDV RNA negative, anti-HD positive patients and also in HDV RNA negative, antibody positive woodchucks may suggest their presence (Smedile *et al.*, 1991; Karayiannis and Monjardino unpublished).

Molecular Pathogenesis

In the study of HDV pathogenesis, two questions have been addressed at the molecular level: how the virus causes damage in the infected hepatocyte, and how it persists, in some cases but not in others, leading to chronic infection. Either the virus or the host or both, will hold the key to these clinically expressed biological phenomena.

Initial studies on the mechanism of HDV-induced cell damage implicated direct toxicity arising from the accumulation of HDAg (Cole *et al.*, 1991; MacNaughton *et al.*, 1990) although the mechanism has since been questioned in a number of reports clearly showing that the presence of HDAg *per se* does not induce cell death. Evidence includes the lack of pathological changes in transplanted livers with only markers of HDV infection (Ottobrelli *et al.*, 1991), the ability to establish viable HDAg-expressing cell lines (Chen *et al.*, 1990; Cheng *et al.*, 1993), the viability of hepatocytes infected in culture (Sureau *et al.*, 1991) and finally the absence of pathological changes in HDAg-expressing transgenic mice. In these transgenic lineages HDAg expression occurred in the presence of HDV replication in one study (Polo *et al.*, 1995) but not in the other (Guilhot *et al.*, 1994) where only HDAg was expressed, both in the presence or absence of HBsAg. Information from such transgenics must, however, be interpreted with some circumspection since they noticeably fail to mimic natural infection. This was illustrated by preferential HDV replication in skeletal muscle with no HDV RNA editing in one case (Polo *et al.*,1995) and by the absence of replication of either virus, HDV and HBV in the other (Guilhot *et al.*, 1994).

Although when taken together the observations so far seem to concur in ruling out the cytotoxicity of HDAg the possibility that the virus may be cytopathic in the presence of replicating HVB has yet to be conclusively disproved. In fact, both the increased pathogenicity of an isolate from the Amazonian region and low pathogenicity of a virus isolated from the island of Archangelou have recently been tentatively linked to genome sequence diversity (Casey *et al.*, 1993; Yang *et al.*, 1995).

An alternative pathogenetic mechanism, i.e., the immune-mediated hepatocyte damage, has also been implicated but its significance has still to be convincingly documented in experiments similar to those carried out for HBV where passive tranfer of lymphocytes from a donor with active hepatitis induced pathological changes in the HBV-expressing liver of transgenic mice indistinguishable from those observed during natural infection.

The second question in HDV-induced molecular pathogenesis, i.e. the characterization of the molecular basis for viral persistence, has also been addressed experimentally in recent years.

Two different scenarios occur which lead to the termination of HDV infection and presumed eradication of the virus. In the first, HBV and HDV co-infect a host and the dual infection is resolved as a result of effective neutralization of HBV, which presumably disposes of both free HBV and HDV, and of HBV-infected and HBV+HDV infected hepatocytes. As to hepatocytes that might have been infected with HDV alone the absence of HBsAg will prevent the assembly and export of new HDV particles and dissemination of HDV infection will thus be arrested. Clearance of those infected cells will presumably occur as a result of specific cytotoxic T-cell lysis or ultimately through natural hepatocyte death and renewal.

The second scenario involves the superinfection of a host already chronically infected with HBV. In such a setting HDV infection can be either self-limiting or more commonly run a chronic course, and rarely in the former case will it lead to resolution of the co-existent HBV infection.

Several studies have attempted to elucidate the molecular basis of HDV persistence, both in humans and in woodchucks chronically infected with WHV. The latter system offers the advantage that HDV superinfection can be precisely timed and that sequential analysis of events relating to the virus infection and ensuing immune response can be undertaken. Results from such analysis in self-limiting and persistent infections arising from the inoculation of the same virus preparation indicate that HDV- editing efficiency, which depends on host as well as viral factors, will ultimately determine the course of the infection (Yang *et al.*, 1995a).

HDV Infection in WHV-Infected Woodchucks

Whilst HDV does not normally infect woodchucks which are naturally infected with the woodchuck hepatitis virus this was nonetheless achieved in the laboratory. The infectious human serum in this instance appears to have overcome the restricted host range (higher primates) normally displayed by the human hepatitis B Virus (Ponzetto, 1984). Once inside the WHV-infected hepatocyte, the virus was able to replicate and acquire a WHVsAg envelope essential for cell release and subsequent propagation. The woodchuck-adapted HDV model has since been of value in studying the pathogenicity of HDV and the natural immune control of the infection.

The Disease

Natural History

HDV infection can occur either as a result of simultaneous infection with HBV or after infection of a previously HBV-infected carrier. In both cases the disease tends to become more severe as a result of HDV infection with an increased occurrence of both fulminant hepatitis and of an accelerated course of the underlying HBV chronic disease towards cirrhosis and ultimately liver failure.

Both HBeAg positive and HBeAg negative (anti-HBe positive) HBV chronic carriers can be superinfected with HDV which normally causes a transitory inhibition of co-existing HBV replication. After some time this inhibitory effect is overcome and high levels of replication of both viruses have been described in a group of patients (Saldanha *et al.*, 1989) and associated with more aggressive disease (Smedile *et al.*, 1991).

Serological findings distinguish acute from chronic hepatititis. In acute infection, an early phase of antigenaemia is followed by raised IgM antibody which lasts 2–3 weeks and which in turn is followed by the appearance of IgG, whereas in chronic infection both IgM and IgG co-exist throughout the infection. The co-existence of Ig anti-HD and documented viraemia suggests that the antibody is not protective. This appears to be confirmed by the finding that immunisation of woodchucks with genetically-engineered antigen containing the major epitope (aa 13–76) did not prevent subsequent infection when the animals were challenged with virus (Karayiannis *et al.*, 1990). In contrast with the controls, the infection did not resolve and followed a chronic course which was documented for over a period of more than two years. By contrast, partial protection has been reported in woodchucks after immunisation with recombinant vaccinia virus expressing the full antigen (Karayiannis *et al.*, 1993). Because in this system the antigen is processed by the infected cell in a HLA-I restricted fashion which leads to cytotoxic T cell recognition, the preeminence of the cellular immune response in HDV control is presently favoured. This appears to receive support from a recent study where the role of CD4$^+$-T cell mediated immunity has been described (Nisini *et al.*, 1997).

HDV infection is usually asociated with HBs antigenaemia and is thought not to persist in its absence. The virus may enter liver cells and replication may take place in the absence of HBV but infection is unlikely to spread to other cells in the absence of a HBsAg coat which mediates the hepatotropism of the virus and also protects its genome.

The specific diagnosis of HDV infection involves the detection of delta antigen in liver biopsy specimens or in serum and of serum anti-HD. In HBV/HDV co-infection, both or either IgM and IgG anti-HD can be raised in the presence or absence of HDAg. IgM is found soon after the onset of the disease and disappears later whereas IgG is found during convalescence. In chronic infection both classes of Ig are found throughout the disease with detectable HDAg in liver (or serum) in a proportion of the cases depending on the stage of the disease.

Molecular hybridisation using cDNA or single-stranded RNA probes has provided a sensitive and non-invasive assay for the presence of the virus. It has now been used in several laboratories and the presence of RNA in serum was shown to correlate well with the presence of HDAg in the liver. The assay remains however somewhat cumbersome and restricted to research laboratories. Recently polymerase chain reaction (PCR) amplification has been introduced for detection of HDV RNA with a limit of detection down to about 10 fentograms compared to 1pcg for dot-blot hybridisation.

HDV hepatitis is recognised by the detection of HDAg in the nuclei of hepatocytes. In chronic HDV infection marked parenchymal and portal inflammatory, changes have been described, but no distinctive histological features of delta hepatitis have been identified.

Epidemiology

The modes of transmission of HDV are similar to those of HBV as it occurs mainly via the parenteral route through contact with blood or blood derived products. Vertical transmission is a rare occurence. Serologic studies have shown that HDV is found all over the world although its prevalence appears not to follow the prevalence of HBV and to show an epidemiology of its own.

Very high rates of HDV prevalence have been reported in the northern part of South America where, in Colombia and amongst the Yncpa Indians of Venezuela, anti-HD positivity is found in up to 80% of HBsAg carriers and is associated with high rates of fulminant hepatitis. Similarly high rates have been reported from central Africa, Asiatic Russia and Romania and rates of up to 66% in HBsAg carriers with chronic liver disease. In Italy rates of anti-HD of up to 50% are found amongst chronic carriers of HBsAg whereas lower incidences are found in other mediterranean countries like Greece and Spain. However as recent reports indicate a dramatic decline in HBV infection in some of these countries (D'Amelio *et al.*, 1992) a corresponding decrease in HDV prevalence can be anticipated.

In Northern Europe and North America the incidence of anti-HD is very low and almost entirely confined to intravenous drug abusers in whom prevalence can be high (30%–75%). In the Far East (China and Japan) where HBV is quite prevalent, antibodies against HDV are very rare in healthy carriers and only about 2%–3% amongst patients with chronic diseases. The same is observed amongst other groups with high incidence of HBV like the inhabitants of Southern Africa and the Alaskan Eskimos. The high incidence of HDV infection in Romania (80%) contrasts with prevailing rates of 7.1% and 8.6% in neighbouring countries as do prevalence rates between northern and southern Kenya. These differences appear to reflect the ongoing spread of the infection as has been shown in Sweden, where no serologic evidence of HDV was found before 1973, and where 80% of drug addicts are now anti-HD positive. Homosexual carriers of HBsAg show a striking low frequency of anti-HD which suggests a different mode of transmission or a more limited spread when compared with drug addicts.

Prevention

It has been estimated that there are about 300 million HBV carriers around the world and that at least 5%, or 15 million, are infected with HDV. In spite of a trend of decreasing HBV incidence recorded in recent years in a number of countries HDV infection still is, and will be for many years, a serious health problem in some regions of the world and the development of a vaccine for the prevention of HDV superinfection of HBV carriers should be considered.

Although our knowledge of the natural immunity against HDV is still fragmentary the evidence available indicates that there is no neutralising B-cell mediated response mounted against HDAg, as one might have predicted from its location inside the virion, and that T-cell mediated immunity may play a more important role. A more detailed understanding of the natural mechanism of virus neutralization is required before vaccine design can be contemplated.

Treatment

Several groups have shown that lymphoblastoid and recombinant α-interferon will inhibit HDV replication, at least during the course of the treatment. Total eradication of the virus is however rarely if ever achieved and the virus re-appears following withdrawal of therapy. Other drugs like Ribavirin have been tried but with no beneficial effect and foscarnet and acyclovir appear to enhance HDV replication (Rosina *et al.*, 1994).

Recent reports have indicated that orthopic liver transplants achieve a lasting clearance of HDV in a minority of cases.

Genetic approaches to HDV therapy based on the inhibitory effect of the large HDV antigen on HDV replication remain an unexplored possibility.

References

Bergman, K.F., Cote, P.L., Moriarty, A. and Gerin, J. *J. Immunology* 1989; **143**: 3714–3721).

Bichko, V.V. and Taylor, J.M. *J. Virol.* 1996; **70**: 9064–9070.

Bonino, F., Hoyer, B., Shih, J.W-K. *et al. Infect. Immun.* 1984; **43**: 1000–10005.

Bonino, F., Heermannn, K.H., Rizzetto, M. and Gerlich, W. *J. Virol.* 1986; **58**: 945–950.

Brazas, R. and Ganem, D. *Science* 1996; **274**: 90–94.

Branch, A.D., Benenfield, B.J., Baroudy, B.M. *et al. Science* 1989; **243**: 649–652.

Casey, J.L., Bergmann, K.F., Brown, T.L. and Gerin, J. *PNAS* 1992; **89**: 7149–7153.

Casey, J.L., Brown, T.L., Colan, E.J. *et al. PNAS* 1993; **90**: 9016–9020.

Chao, Y-C., Chang, M-F., Gust, I. and Lai, M.M.C. *Virology* 1990; **178**: 384–392.

Chao, Y-C., Lee, C-M., Tang, H-S. *et al. Hepatology* 1991; **13**: 345–352.

Chang, F-L., Chen, P-J., Tu, S-J. *et al. PNAS* 1991; **88**: 8490–8494.

Chen, P-J., Kalpana, G., Goldberg, J. *et al. PNAS* 1986; **83**: 8774–8778.

Chen, P-J., Kuo, M.Y-P., Chen, M-L. *et al. PNAS* 1990; **87**: 5253–5257.

Cheng, D., Yang, A. Thomas, H. and Monjardino, J. In *Hepatitis Delta Virus* (eds.) Hadzyiannis, S., Bonino, F. and Taylor, J. *Wiley-Liss, Inc.* 1993 p149–153.

Cole, S.M., Gowans, E.J., MacNaughton, T.B. *et al. Hepatology* 1991; **13**: 845–851.

Cunha, C., Monjardino, J., Cheng, D. *et al. RNA* 1998; *in press.*

D'Amelio, R., Matricardi, P.M., Biselli, R. *et al. Am. J. Epid.* 1992; **135**: 1012–1018

Dourakis, S., Karayiannis, P., Goldin, R. *et al. Hepatology* 1991; **14**: 534–539.

Fu, T-B. and Taylor, J. *J. Virol.* 1993; **67**: 6965–6972.

Glenn, J. and White, J.M. *J. Virol.* 1991; **65**: 2357–2361.

Glenn, J., Watson, J.A., Havel, C.M. and White, J.M. *Science* 1992; **256**: 1331–1333.

Guilhot, S., Huang, S-N., Xia,Y-P. *et al. J. Virol.* 1994; **68**: 1052–1058.

Hwang, S.B., Lee, C.Z. and Lai, M.M.C. *Virology* 1992; **190**: 413–422.

Imazeki, F., Omato, M. and Ohto, M. *J. Virol.* 1990; **64**: 5594–5599.

Karayiannis, P., Saldanha, J.A., Monjardino, J. *et al. Hepatology* 1990; **12**: 1125–1128.

Karayiannis, P., Saldanha, J.A., Jackson, A. *et al. J. Med. Virol.* 1993; **41**: 210–214.

Kimura, M. *Nature* 1968; **217**: 624–626.

Kos, A., Dijkema, R., Arnberg, A.C. *et al. Nature* 1986; **323**: 558–560.

Kos, T., Molijn, A., van Doorn, L-J. *et al. J. Med. Virol.* 1991; **34**: 268–279.

Krushkal, J. and Lee, W-H. *J. Mol. Evol.* 1995; **41**: 721–726.

Kuo, M.Y-P., Goldberg, J., Coates, L. *et al. J. Virol.* 1988a; **62**: 1855–1861.

Kuo, M.Y-P., Sharmeen, L., Dieter-Gotlieb, G. and Taylor, J. *J. Virol.* 1988b; **62**: 4439–4444.

Kuo, M.Y-P., Chao, M. and Taylor, J. *J. Virol.* 1989; **63**: 1945–1950.

Lazinski, D.W. and Taylor, J. *J. Virol.* 1993; **67**: 2672–2680.

Lee, C-M., Bih, F.Y., Chao, Y-C. *et al. Virology* 1992; **188**: 265–273.

Lin, J-H., Chang, M-F. and Baker, S.C. *et al. J. Virol.* 1990; **64**: 4051–4058.

MacNaughton, T.B., Gowans, E.J., Jilbert, A.R. and Burrell, C.J. *Virology* 1990; **177**: 692–698

Makino, S., Chang, M.F., Shieh, C.K. *et al. Nature* 1987; **329**: 343–346.

McNair, A.N., Cheng, D., Monjardino, J. *et al. J. Gen. Virol.* 1994; **75**: 1371–1378.

Melcher, T., Maas, S., Herb, A. *et al. Nature* 1996; **379**: 460–463.

Negro, F., Cobra, B.E., Forzani, B. *et al. J. Virol.* 1989; **63**: 1612–1618.

Netter, H.J., Wu, T-T., Bockol, M. *et al. J. Virol.* 1995; **69**: 1687–1692.

Nisini, R., Pasoh, M., Accapezzato, D. *et al. J. Virol.* 1997; **71**:2241–2257.

Ottobrelli, A., Marzano, A., Smedile, A. *et al. Gastroenterology* 1991; **101**: 1649–1655.

Polo, J.M., Jeng, K.S., Lim, B. *et al. J. Virol.* 1995; **69**: 4880–4887.

Polson, A.G., Bass, B.L. and Casey, J.L. *Nature* 1996; **380**: 454–456.

Ponzetto, A., Cote, P.J. , Popper, H. *et al. PNAS* 1984; **81**: 2208–2212.

Riesner, D. and Gross, H.G. *Ann. Rev. Biochem.* 1985; **54**: 531–564.

Rizzetto, M., Canese, M.G., Arico, S. *et al. Gut* 1977; 997–1003.

Rizzetto, M., Hoyer, B., Canese, M.G. *et al. PNAS* 1980; **77**: 6124–6128.

Robertson, H.D., Manche, L. and Mathews, M.B. *J. Virol.* 1996; **70**: 5611–5617.

Rosina, F. *Antiv. Res.* 1994; 165–174.

Ryu, W-S., Bayer, M. and Taylor, J. *J. Virol.* 1992; **66**: 2310–2315.

Ryu, W-S., Netter, H.J., Bayer, M. and Taylor, J. *J. Virol.* 1993; 3281–3287.

Saldanha, J., di Blasi, F., Blas, C. *et al. J. Hepatol.* 1989; 23–28.

Saldanha, JA, Homer, E., Goldin, R. *et al. J. Gen. Virol.* 1990; **71**: 471–475.

Saldanha, J.A., Thomas, H.C. and Monjardino, J. *J. Gen. Virol.* 1990; **71**: 1603–1606.

Sharmeen, L., Kuo, M.Y-P., Dieter-Gotlieb, B. and Taylor, J. *J. Virol.* 1988; **62**: 2674–2679.

Smedile, A., Rosina, F., Saracco, G. *et al. Hepatol.* 1991; **13**: 413–416.

Sureau, C., Jacob, J.R., Eichberg, J.W. and Landford, R.E. *J. Virol.* 1991; **65**: 4292–4297.

Sureau, C., Moriarty, A.M., Thornton, G.B. and Landford, R.E. *J. Virol.* 1992; **66**: 1241–1245.

Wang, H-W., Chen, P-J., Wu, J-C. *et al. J. Virol.* 1991; **65**: 6630–6636.

Wang, K.S., Choo, Q.L., Weiner, A.J., *et al. Nature* 1986; **323**: 508–514.

Wu, J-C., Chen, C-M., Sheen,I -J. *et al. Hepatology* 1995; **22**: 1656–1660.

Wu, H-N. and Lai, M.M. *Science* 1989; **243**: 652–654.

Yang, A., Papaioannou, C., Hadzyiannis, S. *et al. J. Med. Virol.* 1995; **47**: 113–119.

Yang, A., Karayiannis, P., Thomas, H.C. and Monjardino, J. *J. Gen. Virol.* 1995; **76**: 3071–3078.

Xia, Y-P. and Lai, M.M.C. *J. Virol.* 1992; **66**: 914–921.

PART 4

HEPATITIS C VIRUS

History

In the early seventies, after the agents causing infectious hepatititis (HAV) and serum hepatitis (HBV) had been identified and preventive measures against these viruses implemented, it became apparent that there was still a residual group of infectious hepatitis. This group was known as non-A, non-B hepatitis and included cases associated with previous contact with blood or blood products, others where no such link could be established, and also an 'epidemic' type which appeared to be transmitted by the foecal-oral route and which caused a particularly severe illness in pregnant women.

After a decade and a half of unsuccessful attempts to find the agent(s) in many laboratories throughout the world a non-A, non-B virus was finally identified in 1989, designated Hepatitis C Virus (HCV), and later shown to be the agent responsible for more than 90% of cases of non-A, non-B post-transfusion hepatitis in the Western World. HCV was discovered by workers at Chiron Corporation who succeeded in identifying an epitope-bearing clone from a bacteriophage expressing library made from a non-A, non-B serum-derived nucleic acid preparation (Choo, Q-L. et al., 1989). Once the original clone was confirmed it was possible to extend the sequence to the rest of the genome and also to identify other epitopes amongst the newly-cloned genes. Based on this information a first generation of HCV antibody assays were developed for diagnosis and large scale blood screening.

Of the original group of non-A, non-B hepatitis, HCV was found to account for the large majority of non-A, non-B post-transfusion or blood-associated hepatitis cases and also of those in whom no risk factors could be identified (community acquired or sporadic). The 'epidemic' form has since been shown to be caused by a distinct virus, hepatitis E virus (HEV), which is described in

Part 5. Another human agent, HGV or GBVC, which has been recently described, has yet to be conclusively associated with any human pathology (see Part 6).

The Virus

Classification: New Genus of Flaviviridae

HCV is a positive strand RNA virus that is distantly related to flaviviruses and pestiviruses. Sequence homology between HCV and the two genera is low and restricted to the 5' Non-coding region of pestiviruses whereas its overall genomic organization both in terms of gene ordering within the genome and gene size is closer to that of flaviviruses. HCV represents a new genus within the Flaviviridae family.

The Virion

The virion has not been adequately characterized. This is due to the small amounts of virus available from infected serum and to the present inability to grow the virus in cultured cells.

The reported particle density varies between about 1.10 g/cm^3 or less (Bradley *et al.*, 1991; Miyamoto *et al.*, 1992; Hijikata *et al.*, 1993); 1.115 g/cm^3 (Takahashi *et al.*, 1992); 1.15 g/cm^3 (Kaito *et al.*, 1994); and 1.14–1.18 g/cm^3 (Abe *et al.*, 1989). The differences in density are not immediately explainable but some studies indicate that a lighter lipoprotein-rich fraction may shift to the denser fraction after detergent treatment which removes the outer coat and releases the nucleocapsid cores. Similar low-density particles have also been described in the case of pestiviruses, bovine diarrhea virus has a density of 1.09–1.16 g/cm^3, but not in flaviviruses which have higher densities around 1.20 g/cm^3 (reviewed by Horzinek, 1989). As to the size of the particle, some studies refer a smaller size of 30–60 nm or 30–38 nm (He *et al.*, 1987; Yuasa *et al.*, 1991) whereas others agree on a diameter of about 60 nm which suggests that the smaller particles are probably nucleocapsid cores. Size determinations involved the use of filters of graded pore size and the filtrate was either monitored for infectivity by inoculation of chimpanzees, or for viral RNA by PCR amplification. Recently, two electron microscopy studies have described serum-derived particles of diameters between 55–65 nm (Kaito *et al.*, 1994) corresponding to a fairly heterogeneous spread of particle densities and between 60–75 nm for particles of density 1.10 g/cm^3 (Prince, A.M. *et al.*, 1996). Two fractions of the virus were reported in serum, one which appears to be made up of virus-immunoglobulin complexes and which is presumed non-infectious and another, a lighter fraction, which is infectious and where the virus is associated with lipoproteins. These observations await confirmation with the purified fractions once infectivity assays become available.

The Genome

Many full sequences (but for the 3′ NC region)of the HCV genome have now been reported from a variety of countries throughout the world (Kaito *et al.*, 1990; Takamizawa *et al.*, 1991; Okamoto *et al.*, 1991; Choo *et al.*, 1991; Chen *et al.*, 1992).

HCV RNA is single stranded, about 9600 nucleotides long and of positive polarity, and contains a long Open Reading Frame encoding 3011 amino acids which spans almost the whole length of the genome (Fig. 18). The long ORF is flanked by two non-coding regions, one at the 5′ end and 341 nucleotides long which is thought to contain essential regulatory sequences and is well-conserved amongst isolates, and another located at the 3′ end which appears to be more variable both in length and in primary structure. The latter was initially thought to terminate in a homopolymeric tail of poly U but is now known to extend beyond the poly U stretch by an additional conserved sequence of 98 nucleotides (Tanaka *et al.*, 1995).

Based on the analogies with pesti- and flaviviruses and hydrophilicity/hydrophobicity predictions of putative gene products, HCV structural genes could be mapped to the 5′ end of the ORF whereas the non-structural genes followed downstream from E2 to the 3'end of the ORF (Fig. 18). Sequential processing of gene products from the translated polyprotein and definition of the boundaries for the different genes have been achieved from *in vitro* translation of synthetic viral RNA templates; from transfection of cells with DNA copies of HCV genes and from infection of cells with either recombinant vaccinia or baculo viruses expressing HCV genes.

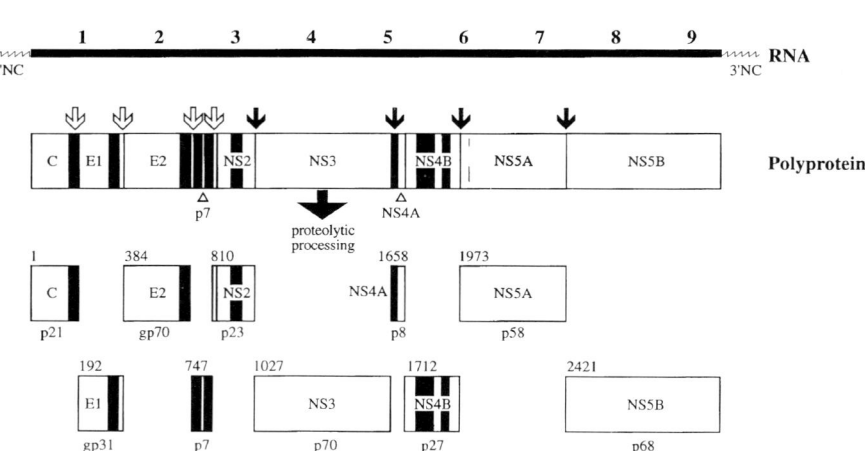

Fig. 18   HCV Genome and proposed gene boundaries. White arrows: Cleavage points by host protease(s); Black arrows: Cleavage points by HCV proteases. Black areas: hydrophobic regions. Figures refer to the aminoterminal amino acid for each protein. *p* or *gp* followed by a figure indicate protein or glycoprotein and respective size in kilodaltons.

Genotypes

HCV genome sequence variation has been reported and six main genotypes, each comprising different subtypes, have been described (Simmonds *et al.*, 1993). The different genotypes differ from each other by approximately 30% over the whole genome. Type 1 is the most prevalent type in most countries, including those in Western Europe, North America, the Indian Sub Continent and the Far East, as well as in Australia. Southern Africa shows a predominance of type 5 but type 4 has also been reported in this region, becomes common in Central Africa and is the most prevalent type in Egypt. Type 2 is not uncommon in the Far East and has been also reported in Northern Europe and Type 3 is common in the Indian Sub-Continent. Finally type 6 is found in Hong-Kong and also in Thailand and neighbouring countries. Of the two major subtypes of Group 1, 1a and 1b, predominance of one over the other also varies across the world with 1b predominanting in the Far East and also in some European countries and 1a being the commonest in the USA and other Western European Countries.

Sequence diversity varies throughout the genome with 5' NC, NS4B and core being the most conserved and E2 and E1 the most variable. The additional X sequence in 3' NC region recently described also appears to be well preserved.

Genotyping between the 6 main groups has been carried out by restriction digestion, reverse-hybridization (to cold specific immobilized oligos), and sequencing of RT-PCR amplified cDNA and also by serotyping to group-specific peptides. Although the latter is indirect and may reflect a previous rather than a current infection this has not proved to be a significant problem in the studies published sofar.

5' and 3' Non-Coding Regions

The 5' end non-coding region is 341 nucleotides long and shows significant sequence identity with the 5' leaders of animal pestiviral genomes (Brown *et al.*, 1992). The region shows extensive secondary structure and a stem-loop domain III which in one proposed model, is almost superimposable to similar domains in pestiviruses in spite of low level (47%) primary sequence relatedness (Brown *et al.*, 1992). The 5' NTR also contains two pyrimidine tracts of which one sequence, CCUUUCUUGGA, is complementary to 18S ribosomal RNA and located within the apical loop of the domain III of the stem-loop structure. Such sequences were previously identified in picornavirus 5' NTRs with ribosome landing pads (Pilipenko *et al.*, 1992). The existence of an Internal Ribosome Entry Segment (IRES) in the HCV genome has now been clearly established (Tsukyiama-Kohara *et al.*, 1992), by *in vitro* translation experiments under conditions where capped mRNA are not translated and by using bi-cistronic transcripts, and although its boundaries have not been precisely mapped, genetic-functional dissection based on the introduction of mutations which abolish individual elements of secondary structure has indicated that the stem-loop I and II structures and first polypyrimidine tract sequence (nucleotides 120–130) appear not to be essential for *in vitro* translation, that the 5' hairpin is inhibitory, and that a stem within stem-loop domain III is essential (Yen *et al.*, 1995).

In contrast with the 5' terminus which shows a well conserved sequence, the 3' non-coding region varies considerably in length. Initially thought to terminate in a homopolymeric tail (poly U or poly A) it is now known to extend beyond the poly U stretch by an additional well-conserved sequence of 98 nucleotides (Tanaka *et al.*, 1995). Differences in length between isolates are exclusively due to varying sizes of the sequence between the 3′ end of the ORF and the start of the poly U tail as well as to the variable length of the tail itself. The 3′ NC region has a high content of secondary structure, a feature which explains the technical difficulties encountered in obtaining the full sequence.

The well-preserved 98 nucleotide sequence is thought to have an essential role in viral replication, most probably involving a 'promoter-like' function for the virus-coded RNA-dependent RNA polymerase.

Structural Proteins

The ORF encodes 10 genes of which three are structural: core (C), envelope 1(E1), and envelope 2(E2); and seven are non-structural: NS2, NS3, NS4A, NS4B, NS5A, and NS5B (Fig. 18). An additional sequence p7 has been mapped between E2 and NS2 which in contrast to all other genes, appears not to be co-translationally processed (see below).

Core

The core gene has been mapped to the 5′ end of the ORF based on its hydropathic profile and similarities between HCV and flaviviruses' genomic organization (pestiviruses have a non-structural gene in the corresponding location). The suggested boundaries of the gene are nucleotides 342–914; 342 being the A of the initiation codon AUG and 914 being the third base of the alanine residue at the 191/192 aa signalase cleaving site (Fig.18). The predicted size of the core protein is 22.5 kD and the protein is predominantly hydrophilic and positively charged, with more than 20% lysine plus arginine residues, in line with its predicted interaction with the negatively charged genome. The protein also contains a hydrophobic region at its carboxyl end which would predictably anchor the protein to the endoplasmic reticulum. A protein of 22 kD, in good agreement with its predicted size, has been expressed in cultured cells after transfection of cDNAs containing the putative core gene (Harada *et al.*, 1991; Hijikata *et al.*, 1991; Kumar *et al.*, 1992) and cytoplasmic staining showing core antigen associated with the endoplasmic reticulum is compatible with the anchoring of the protein to the reticulum. Nuclear staining, almost exclusively restricted to the nucleoli, was also reported 7–9 days after transfection (Fig. 19) (Kumar *et al.*, 1992). This observation has since been confirmed (Shih *et al.*, 1993) and the nuclear core polypeptide has been found to be amino co-terminal but shorter (16 kD) than the cytoplasmic form (Lo *et al.*, 1994) (Fig. 20). The shorter form is thought to result from the processing of the anchored 22 kD molecule (Chang *et al.*, 1994) and to be directed to the nucleus via nuclear localization motifs. Although it is not yet known whether HCV core protein migrates to the nucleus during natural HCV infection, a similar finding has been reported during infection with flaviviruses. Processing of HCV core is carried out by a host signalase.

Although little is known about the assembly of HCV recent studies have reported the formation of core homodimers which may be the structural unit for the formation of capsid particles and for the interaction between core and E1, but not E2. Such interaction was shown to involve the hydrophobic regions of the caboxy termini of both core and E1 and is likely to occur in the endoplasmic reticulum membrane.

The core protein is immunogenic and constitutes one of the antigens included in the current HCV antibody assays. Several linear epitopes have been mapped with the use of synthetic peptides in direct binding and in competing assays with patients' sera. Immunodominant epitopes were reported within the first 74 amino-terminal amino acids (Nasoff *et al.*, 1991) and have since been confirmed and more precisely mapped (Sallberg *et al.*, 1992). T cell core epitopes have also been described (see below).

An involvement of HCV core protein in hepatocarcinogenesis has recently been suggested in a report of cooperation with *ras* leading to transformation of primary liver fibroblasts (Ray *et al.*, 1996).

Recombinant HCV core protein has been expressed in a variety of expressing systems, both eucaryotic and procaryotic.

Envelope 1 (E1)

The E1 gene of HCV has been mapped 3′ to the core gene between aas 192 and 383 for HCV1 (Fig. 18). The first amino acid of E1 has been confirmed by microsequencing as Tyr[192] (Hijikata *et al.*, 1991a) which is preceded by a

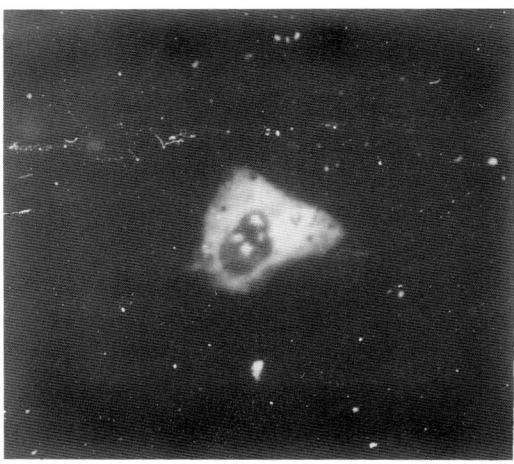

Fig. 19 HCV core protein expression after transfection of Huh-7 human hepatoma cells. Nucleolar accumulation of antigen is seen after 9 days in addition to strong (membrane-associated?) cytoplasmic staining (Kumar *et al.*, 1993).

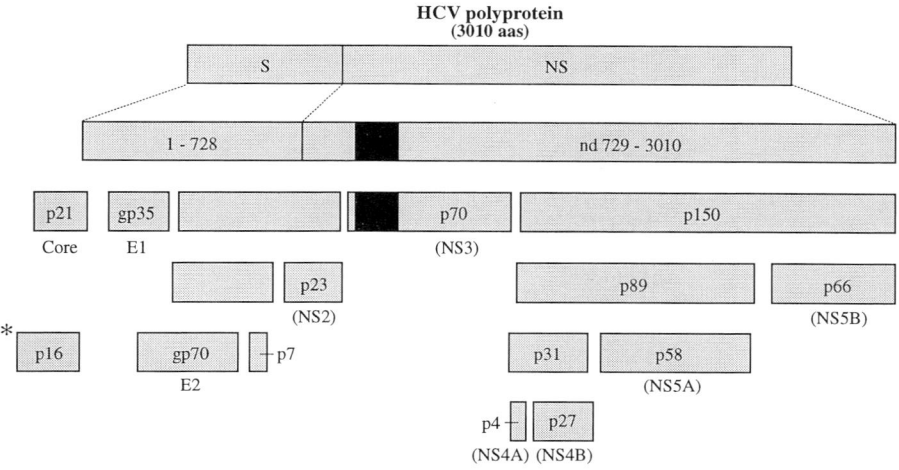

Fig. 20 Proposed sequential processing of HCV polyprotein. Solid black region within p70 indicates position of protease domain. *p16 indicates nuclear form of core polypeptide.

hydrophobic segment containing a site for signalase cleavage. A host signalase is thought to be involved. The predicted molecular weight is around 21 kD but modification through N-linked glycosylation of four of the five potential sites generates a product with a molecular weight of about 35 kD (Fig. 20).

The E1 glycoprotein forms a stable complex with E2 which has been demonstrated by co-immunoprecipitation (Grakoui *et al.*, 1993; Ralston *et al.*, 1993). The E1: E2 complex was shown not to involve disulfide bonding and preliminary sedimentation data suggests a possible trimeric complex of E1-E2 heterodimer (Ralston *et al.*, 1993). The rate of formation of E1: E2 heterodimers was found to be slow when compared with other viruses but this may reflect the slow folding of E1 and/or E2 which requires interactions with appropriate molecular chaperones (Dubuisson *et al.*, 1994). A recent report suggests that the disulfide-dependent intramolecular folding of E2 occurs rapidly but that that of E1 is slow and possibly dependent on a prolonged association with the molecular chaperone calnexin (Dubuisson and Rice, 1996).

Studies where intracellular traffic of E1 and E2 was analyzed revealed that no complex sugars are added to the two glycoproteins, as shown by the absence of H-endoglycosidase resistant forms, and that, in the system used, they are essentially retained in the ER in and do not reach the Golgi. (Dubuisson *et al.*, 1994).

Envelope 2(E2)

A gene which for some time was designated E2/NS1 has now been clearly established as a second envelope glycoprotein or envelope 2. It maps between aas 383 and 729 (Fig. 18). The molecular weight of the gene product is 70 kD. Nine potential N-linked glycosylation sites were identified and are thought to be occupied. The glycoprotein contains twenty cystein residues which are expected to be part of disulfide bridges.

Three carboxy-terminally different E2-containing polypeptides have been identified which reflect the complex processing of this region of the polyprotein between the structural and non-structural genes. The three forms include E2 itself which is presumed to be processed by a host signalase at A^{729} and at the end of the hydrophobic stretch at the carboxyl end, and two E2 precursors, E2-NS2 and E2-p7, which are processed with variable efficiency according to genotype. The latter demonstrate that whereas elsewhere in the genome processing of the polyprotein is co-translational at both the E2/p7 and p7/NS2 junctions it can be posttranslational (Fig. 20)..

The gene shows marked sequence variation between isolates and it also contains regions which show genetic hypervariability during the course of the disease. Two of these have been clearly identified, hypervariable region 1 (HR1) which lies at the amino terminus of E2 between aas 386–411, and HR2 which maps between 474 and 480 (Hijikata *et al.*, 1991b; Weiner *et al.*, 1991). HV1 has been shown to undergo 0.5–1.7 amino acid changes per month and appears to be isolate-specific and hence of great value in epidemiological studies. Comparison of HVR1 published sequences (Lesniewski *et al.*, 1993) revealed that only 4 amino acids, Thr^{385}, Gly^{406}, Gln^{409} and Leu^{413} are preserved with frequencies of 98.2%–100% (with only one, glycine at position 406 found in 100%). The HVR1 appears to be the site of at least one B-cell epitope. It has been suggested that this may be a neutralizing epitope, as reproted for the second envelope protein of pestiviruses, and that its mutability reflects the phenomenon of immune escape and constitutes the molecular basis of viral persistence (see PATHOGENESIS).

P7

This sequence is located between E2 and NS2 and maps between aas 729 and 809 (Selby *et al.*, 1994; Lin *et al.*, 1994) (Fig. 18). At both its amino and carboxyl termini it is thought to be processed by host signalases and cleavage sites have been identified. As indicated before processing in this region is posttranslational and follows proteolytic cleavage at the NS2/NS3 junction by the NS2 protease. The function of p7 is unknown.

Non-Structural Proteins

NS2

The boundaries of NS2 lie between residues 810 and 1027 (Fig. 18). Whereas processing of the three structural proteins at the core/E1 and E1/E2, and E2/p7 and p7/NS2 boundaries is thought to be carried out by host signalases, since it is successfully carried out in the absence of the non-structural genes, the cleavage at the amino end of NS3 (NS2/NS3 boundary) is thought to be carried out by a virus-encoded protease. This protease is thought to be a zinc-dependent metalloprotease which requires, for efficient processing, amino acid sequences upstream of the cleaving site up to 849–923 and also downstream into NS3 between 1137 and 1237, hence overlapping the serine protease domain which maps to the amino end of NS3 (Fig. 18). Residues His-952 and Cys-993 are essential for processing at the 2/3 site (Hijikata *et al.*, 1993b; Grakoui *et al.*, 1993b). This metalloprotease activity can be shown to be distinct from the NS3 protease by the use of mutations which abolish the latter but not the former. In conjunction with a host signalase which cleaves at the N-terminus of NS2 at position 810, both enzymes release the NS2 protein (Fig. 20). Evidence that a host signalase is involved in processing the N-terminus of NS2 comes from the identification of a signalase cleavage site motif and from the observation that replacement of alanine at the putative cleavage site inhibited membrane-dependent processing. Proteolytic cleavage at the NS2/NS3 site is thought to operate in *cis*, i.e. to be autoproteolytic, but a bimolecular reaction can occur *in vitro* with portions of NS2 and NS3 (Reed *et al.*, 1995). Recent evidence has suggested that NS2 protein is a transmembrane protein with a cytosolic N-terminus and a C-terminus located in the lumen of the ER (Santolini *et al.*, 1995). Although no series of hydrophobic amino acids that could act as signal sequences or transmembrane domains could be identified in NS2, there is evidence to suggest that the sequence involved in membrane targetting lies downstream from the cleavage site and that targetting may involve host proteins. It has been proposed that the polyprotein molecule first folds for the functional reconstitution of the autoprotease which performs the autocatalytic cleavage which is then followed by the insertion of NS2 into the ER membrane.

NS3

The NS3 gene product has been mapped between aas 1027 and 1657 (Fig. 18) and the amino terminus has been confirmed by microsequencing of the expressed protein which has a molecular weight of 70 kD. Initial analysis of the putative amino acid sequence detected some homology with flavi and pestiviruses' NS3 and p80 proteases, respectively. Highly conserved amino acids His[1083], Asp[1107] and Ser[1165] were identified as the possible catalytic triad of a serine protease and so was a nucleoside triphosphate-binding helicase domain (Miller and Purcell, 1990; Chambers *et al.*, 1990; Choo *et al.*, 1991). X-ray crystallographic analysis of the NS3 protease domain revealed that it folds as chymotrypsin-like protease with 2 β barrels, confirmed the catalytic triad and a substrate-binding site consistent with cleavage specificity, and identified a structural zinc-bing site. (Love *et al.*, 1996; Kim *et al.*, 1996).

The NS3 protease, which maps within the N-terminal 167 aas of NS3 from aa 1049 to aa 1215, is involved in the processing of the non-structural genes, operating in *cis* to process the carboxyl terminus of NS3 (NS3/NS4A boundary) and in *trans* when processing the NS4A/NS4B, NS4B/NS5A and NS5A/NS5B junctions (Tomei *et al.*, 1993) (Fig. 20).

Further analysis of the cleavage site (between P1 and P1′) has shown that a minimum HCV sequence between P4 and P1′ is required for trans cleavage and that in the P1 position at the N-terminal cleavage sites in NS4B, NS5A and NS5B cysteine is strongly favoured since substitution at this position abolished or inhibited cleavage (Komoda *et al.*, 1994). By contrast threonine is the conserved P1 amino acid at the NS3/NS4A boundary, where cleavage catalyzed by NS3 protease occurs in *cis*, and of the several substitutions at this position, only Arg was associated with strong inhibition. At the P1′ position, requirements do not appear to be as stringent as for P1 in *trans*-mediated cleavage and variable effects of specific substitutions possibly reflect the different assays used (Komoda *et al.*, 1994; Kolykhalov *et al.*, 1994).

The chronological sequence of intermolecular (or in *trans*) proteolytic cleavages has been established as NS5A-NS5B > NS4A-NS4B > NS4B-NS5A, the latter being dependent on the formation of a complex between NS3 protease and NS4A (Bartenschlager *et al.*, 1994; Lin *et al.*, 1993; Tanji *et al.*, 1994) (Fig. 20). The same hierarchy of NS3 substrate processing was recently demonstrated with synthetic decamer peptides and purified E.Coli-expressed NS3 (Steinkuhler *et al.*, 1996).

The study of the enzyme kinetics with the bacterial enzyme has demonstrated its lowest affinity for the NS4B-NS5A- derived peptide and the highest for the NS4A-NS4B-derived peptide although the conversion rate for the latter was 14 times lower than that for NS5A-NS5B peptide. As proposed for other viruses it has been suggested for HCV that the differential proteolysis rates observed at the various sites may, in fact, constitute a regulatory mechanism which is the counterpart of protein synthesis regulation seen in other systems.

NS3 protein has also been shown to transform NIH 3T3 mouse fibroblasts, an effect thought to be associated with its protease activity (Sakamura *et al.*, 1995). The significance of this finding in the context of the high association between HCV chronic infection and hepatocellular carcinoma is not clear.

The pivotal role of NS3 protease in the processing of the NS region of the polyprotein have made this enzyme one of the chosen targets for antivirals presently being developed.

Finally two other activities have been mapped to NS3, the polynucleotide-stimulated triphosphatase and the helicase. As with members of the other two genera of the *Flaviviridae* family, i.e., flaviviruses and pestiviruses, conserved sequence motifs predictive of nucleotide triphosphatase (NTPase)/helicase have been identified in the HCV sequence within the NS3 region. These are the D-E-A-D sequences and proteins containing such motifs, the DEAD protein family, comprise three subfamilies, DEAD, DEAH, and DEXH. In the case of HCV, DEXH motifs can be identified in the carboxy two thirds of NS3. Both helicase and triphosphatase activities have now been expressed and characterized as recombinant proteins (Suzich *et al.*, 1993; Kim *et al.*, 1995). The NTPase activity prefers ATP as substrate and is preferentially stimulated by poly U and the helicase requires ATP and a divalent ion and unwinds duplexes in a $3' \rightarrow 5'$ direction (Hong *et al.*, 1996). Both these activities have been shown not to require the amino terminal sequence of NS3, where the protease maps, although there is no evidence to suggest that these domains exist separately *in vivo*. Recent advances will offer the opportunity of investigating the *in vitro* activities of these enzymes on a natural template so as to gain further insights into their roles in the replicative cycle of HCV.

NS4A

This polypeptide is 54 amino acids long (aas. 1657–1711), has a molecular weight of about 6 kD and contains a hydrophobic middle-portion and a charged, acidic, C-terminus (Fig. 18). The polypeptide has been found to act as a co-factor of NS3 serine protease and to be required for processing at the NS3/NS4A, NS4A/NS4B, NS4B/NS5A but not at the NS5A/NS5B junction.

NS4 is thought to operate in *cis* when processing the NS3/NS4 junction and in *trans* when processing the remainder. The middle amino acids (aa 21–34) have been associated with the stimulatory effect on NS3-mediated proteolysis. The NS4 poplypeptide appears to form a stable complex with NS3 which requires the N-terminal 22 amino acids of the latter and NS3 molecules that lacked these amino acids showed markedly reduced cleavage at the NS4B/NS5A junction. Analysis of X-ray crystal structure of the NS3 protease domain complexed with a synthetic NS4A activator peptide revealed that the interacting NS4A forms a β strand that intercalates into the N-terminal domain β sheet of the core of the enzyme (Kim *et al.*, 1996). The mechanism of NS4A activation has not been completely elucidated. It has been suggested that the role of NS4A is to mediate the membrane association of NS3 (Hijikata *et al.*, 1993c) but in addition to membrane anchoring other proposed roles for NS4A include the suggestion that its complex with NS3 constitutes the active protease or, alternatively, that it acts as a chaperonin to fold correctly the protease domain or to facilitate interaction with the polyprotein. Evidence supporting the latter, i.e., more efficient interaction with substrate, comes from *in vitro* experiments using purified NS3 protease where a 1:1 complex between a NS4 active peptide and protease shows maximum effect on the cleavage of NS4A-NS4B and NS5A-NS5B by increasing k_{cat} (Steinkuhler *et al.*, 1996).

The requirement for a two-component protease appears to be common to all flaviviruses where NS2 is required to complement NS3-mediated proteolysis. In all these cases, however, the formation of a stable complex appears to be essential for all processing steps, a requirement which does not appear to apply in the case of HCV. Rather more similar appears to be the situation with the pestiviruses where the p80 serine protease, the functional equivalent of HCV NS3, appears to require p10 to complement its proteolytic activity.

The importance of NS4A as a co-factor of NS3 protease has also focused a great deal of interest on this small polypeptide as a possible target for new antivirals.

NS4B

The protein corresponds to aas 1712–1972 (Fig. 18) and the expressed protein has a molecular weight of about 27 kD. Processing is effected by the NS3 protease as indicated above with a strict requirement for NS4A for the processing at the carboxyl (NS4B/NS5A) terminus. The function of NS4B is unknown.

NS5A

NS5A maps between aas 1973 and 2419 (Fig. 18) and the expressed protein has a molecular weight of 56 kD. The protein is phosphorylated and a hyperphosphorylated 58 kD form has also been described. Serine residues which constitute the substrate for phosphorylation have been identified (Tanji *et al.*, 1995).

As indicated above NS5A is processed efficiently and early at its carboxyl terminus (NS5A/NS5B boundary) by NS3 protease, possibly without a requirement for NS4A which then becomes essential for the late processing at its amino terminus (NS4B/NS5A). The protein is found in association with the nuclear periplasmic membrane region and it is of unknown function.

NS5B

The protein maps between aas 2420 and 3011 (Fig. 18) and the expressed protein has a molecular weight of 65 kD. It bears a GDD motif at position 2737–2739 which is characteristic of RNA polymerases of RNA viruses. The NS5B RNA polymerase has been expressed both in insect cells infected with recombinant baculovirus and in bacteria transformed with recombinant plasmids. Low activity levels were obtained from partially purified virus of insect cells and from purified and renatured enzyme expressed in bacterial cells that may reflect the instability of the enzyme (Lin *et al.*, 1994; Yuan *et al.*, 1997; Lohmann *et al.*, 1997). The requirement for co-factors, host, viral, or both host and viral, is presently unknown. HCV polymerase expressed in baculovirus system (Lohmann *et al.*, 1997) was shown to copy the whole of HCV genome and was confirmed to be active on a synthetic poly A template and to lack terminal transferase activity as previously reported (Yan *et al.*, 1997).

Growth of HCV in Cell Culture and HCV Replication

Several reports have claimed HCV replication in a variety of cultured cells, including cells of lymphoid lineage, primary hepatocytes and human hepatoma cells; but no system has yet been established for the efficient productive growth of the virus in culture. A similar difficulty has been encountered with HBV and in both cases inability to infect cells in culture may reflect the absence in the tested cell lines, or the rapid loss in the case of primary hepatocytes, of a specific cell receptor required to internalize the virion.

The alternative approach that bypasses such a potential block, namely the transfection into appropriate cells of full length cDNA or *in vitro* synthesized full length RNA of genomic polarity, has also proved unsuccessful in the case of HCV. By contrast both in flaviviruses (dengue, yellow fever, Western enchephalitis virus) and in pestiviruses (hog cholera virus), transfection of permissive cells with viral nucleic acid has successfully initiated infection. In the case for HBV where transfection of recombinant DNA into differentiated human hepatoma cell lines has been extensively used such an approach will still fail to achieve ultimate propagation of the infection. This is due to the inability of released virions to infect non-susceptible cells which limits replication to a single round. Two recent reports have described the successful infection of chimpanzees after intra-hepatic injection of full-length synthetic HCV RNA (Kolykhalov *et al.*, 1997; Yanagi *et al.*, 1997).

HCV Replication

HCV replication is thought to involve the synthesis of an RNA strand of antigenomic polarity (minus) that will generate a duplex replicative form and this double-stranded RNA will in turn become the template for the synthesis of progeny plus strands.

Initial reports describing the presence of the minus strand replicative intermediate in the liver and lymphocytes using minus strand specific primers for cDNA synthesis were later questioned when it was shown that unprimed cDNA synthesis and subsequent PCR amplification could occur when total tissue RNA was used as substrate. Subsequent modifications introduced to the RT-PCR amplification procedure have now made it specific for minus strand detection (Lanford *et al.*, 1995). Application of such procedures has confirmed HCV replication in the liver but not in peripheral lymphocytes, which was reported as a site of HCV replication in several studies, suggesting that the detection of HCV RNA in those cells may be the result of virus adsorbed to the lymphocyte cell membrane or of abortive infection.

The situation is not any clearer in relation to viral proteins as indirect markers of viral replication. Here again no universal reagents are yet available for HCV protein detection in spite of claims by several groups that antibodies can be used to immunostain HCV proteins in liver sections. The contrast between reproducibly strong staining of HCV recombinant proteins over expressed in eucaryotic cells, and the low, inconsistent, and often undetectable levels during natural infection, when using either patient sera or sera raised against recombinant proteins, clearly points to very low levels of viral expression. This most likely reflects very low levels of viral replication, independently confirmed by undetectable or almost undetectable amounts of HCV RNA in infected liver by *in situ* or Northern blot hybridization and ultimately by the low blood virus levels during infection.

Because RT-PCR of HCV RNA is the only method available to detect the presence of HCV in blood and because of the intrinsically non-quantitative nature of this assay further modifications had to be introduced in order to measure circulating virus levels. Some of the methods presently available are based on the measurement of the unknown sample against known amounts of a distinct (either restriction site-bearing or of different size) competitive template, which is amplified in the same tube and with the same primers. Of the two competitive templates, the type distiguishable by size appears more reliable (if only large or small enough to be clearly identified) than the type bearing a restriction site as the latter is dependent on the subsequent full restriction of the product. Another method for HCV RNA quantitation, i.e., end point amplification of cDNA, is carried out after serial dilution of cDNA until no amplification is obtained and calculations depend on assumptions concerning efficiencies of cDNA synthesis and of PCR amplification. Finally a non PCR-based method is also available where a hybridization-mediated amplification is achieved by a series of specific probes involved in the anchoring, the targetting and the amplification of the HCV RNA ultimately generating a signal which is proportional to the amount of RNA present.

Although some of these assays have lower sensitivities (high cut-off level), of more than 10^5 genomes/ml, results from a variety of studies appear to concur that viraemias range between 5×10^2 and 10^7 genomes/ml with an average of 10^4 to 10^5 and only exceptionally exceeding 10^7 genomes/ml. It is worth pointing out that such assays as those described are not dependent on the presence of the full genome or on its biological viability and that marked discrepancies between determined viraemias and infectivity assays in chimpanzees have been reported that are likely to reflect the presence of large numbers of defective molecules.

Detailed studies of viral replication at the molecular level have been hampered by the absence of a cell culture system which can support efficient viral propagation and by the low levels of HCV replication in infected livers. An alternative *in vitro* approach is dependent on the availability of the purified RNA polymerase and of a synthetic full length HCV genomic RNA template which includes the additional 3′ Non-Coding sequence recently described (Tanaka *et al.*, 1995). Both these requirements appear now to have been met and new information in this area is eagerly awaited.

The Disease

Infection with HCV occurs by the parenteral route and the virus is normally transmitted in blood or blood products. As a result those infected often have a history of blood transfusion or of administration of factor VIII, gamma globulin, anti-D antibody for prevention of Rhesus incompatibility, etc. Intravenous drug abuse with sharing of needles is also a common source of infection and is now the major identifiable risk in countries where blood donations are routinely screened for HCV antibodies. Infection through sexual contact is infrequent as is transmission within the family (horizontal) or between mother and child (vertical).In many cases , up to nearly 40%, the source of infection is not identified. This is partly explained by the long asymptomatic period of the disease and the difficulty in recalling high risk procedures which might have occurred 10–20 years previously. Inadequate sterilisation at the time of mass vaccination or therapeutic campaigns in some countries may also have accounted for a number of epidemics.

The Natural History of the Disease

Acute infection occurs after an incubation period of about 6–10 weeks, depending on the size of the inoculum, and is often asymptomatic. HCV RNA is the first detectable marker, preceding seroconversion by several weeks, followed by the rise in liver enzymes. Antibodies follow, with anti-core normally preceding antibodies against NS3, NS4 or NS5 antigens. Evolution to chronicity occurs in a majority of cases, possibly up to 80%, and both RNA and antibodies remain detectable throughout the course of the disease. Evolution to cirrhosis and liver failure occurs in about 50% of the cases. The remainder include cases of persistent chronic infection with minimal liver damage and a small group of 'healthy' carriers with normal liver biochemistry and histology. Evolution to hepatocellular carcinoma is a common outcome of HCV associated cirrhosis (Colombo *et al.*, 1989; Saiyo *et al.*, 1990).

Protective immunity against HCV infection is still poorly understood. None of the anti-HCV antibodies commercially detected correlate with virus neutralization as demonstrated by the co-existing viraemia. Chimpanzees which have been infected with a defined inoculum show no lasting immunity as shown by the observation that a challenge with a different or even the same inoculum results in an acute hepatitis almost indistinguishable from the original illness (Farci *et al.*, 1992). In extrapolating these findings to the human infection it is important however to note that in the chimpanzees (by contrast with humans), HCV antibody responses, as measured by the commercial assays, tend to be short-lived and that so far in humans almost no documented cases of HCV re-infection have been reported. Convincing evidence to support such ocurrences of sequential rather than mixed infections with two distinct viruses is however intrinsically difficult to obtain in view of the subclinical nature of the acute infection; and the high mutation rate of the virus RNA compounds this difficulty when trying to distinguish an independent re-infecting event from the reactivation of a previous infection.

Indirect evidence pointing to the existence of a changing neutralizing B epitope in the hypervariable region of E2 has accumulated as a result of a number of studies. Evasion of the immune response against such a putative neutralizing epitope in the large envelope glycoprotein has been favoured as the mechanism of virus persistence (Weiner *et al.*, 1992; Kato *et al.*, 1992; van Doorn *et al.*, 1995). The observation that sequence variation in HVR1 is abolished or significantly depressed in agammaglobulinaemic (Kumar *et al.*, 1994) and therapeutically immuno-suppressed patients with actively replicating HCV also appears to support this hypothesis. In this context the fact that such a neutralizing antibody response has not been identified in patients clearing the infection could be explained by the non display of the relevant native epitopes by the purified recombinant antigen preparations so far used in antibody assays.

Just as with B cell responses, studies attempting to identify a specific HCV-neutralizing T-cell mediated response have so far been unsuccessful. Whereas definite proliferative responses, involving lymphocyte T helper cells, against a variety of epitopes in both structural and non-structural recombinant HCV proteins or synthetic peptides have been described none have so far been shown to be specifically associated with resolution of the infection. Significant differences were nonetheless reported between cases where the infection is either undetectable or follows a mild course, and actively infected patients. These relate to the intensity and amplitude (number of positive responses to different antigens) of the T-cell responses, which are clearly reduced in the actively infected group (Lechmann *et al.*, 1996; Missale *et al.*, 1996).

In parallel studies concerning cytotoxic T-lymphocytes, almost all chronic patients showed a response to at least one of the epitopes in core, NS3 or NS4. In spite of the observation that the response was strongest in patients with lower HCV RNA levels, who also showed low serum transaminase levels, a causal relationship is still not established and both findings may independently reflect the changes in T helper cells described above (Rehermann *et al.*, 1996).

The role of host factors, like HLA antigens , and their association either with susceptibility to infection or with chronic carriage is also currently being analysed.

Pathogenesis

The pathological findings reported in hepatitis due to HCV are similar to those found in other viral hepatitis and no specific features characteristic of this type of hepatitis have been described. Some findings are however more frequent in hepatitis C. The general picture is one of chronic inflammation and fibrosis of the portal tracts and of varying degrees of cell necrosis. The more distinctive features include the frequent accumulation of lymphocytes, sometimes forming follicles with germinal centers, and alterations of the bile ducts which show infiltration by lymphocytes and degeneration of the bile duct epithelial cells. The accumulation of fat (steatosis) in hepatocytes is also a more common feature (about 50%) in hepatitis C than in other forms of hepatitis and has been suggested to be the result of the cytopathic effect of the virus.

Although EM changes have not been described in humans cytoplasmic tubular structures were observed in thin-section electron micrographs of infected liver in early studies of non-A, non-B infected chimpanzees and have been subsequently described in confirmed HCV cases.

Epidemiology

The number of carriers of HCV worldwide has been estimated at about 200 million.

The overall average carriage rate is about 1% but it varies between countries, ranging in blood donors between less than 1% in Northern Europe (0.003% in Sweden) and North America and 17% in Egypt. Intermediate rates of around 1–4% are found in Southern European and Middle Eastern countries. The virus does not appear to be very prevalent in the Indian sub-continent where carriage rates are between 0.4–1.5% but is very prevalent in the Far East, particularly in Japan where high endemic areas with carriage rates of more than 10% have been identified but not so prevalent in China and South Vietnam (but highly prevalent in the North) and even less in Taiwan (less than 1%). African countries show varying carriage rates between less than 1% in Kenya and in Mozambican refugees in Swaziland and up to more than 10% in other areas depending on geographic location and ethnic background (with Pygmies showing consistently lower carriage rates than Bantus). Finally in South America both in indigenous and urban populations carriage rates do not appear to exceed 1.5%.

Seropositivity was shown to be more common in men than in women and to increase markedly with age reaching a peak in the third decade and then decreasing in older groups. Hypertransfused patients like hemophiliacs or hemodialysed patients show very high carriage rates as do carriers of other viruses like HIV and HBV which are also transmitted by the parenteral route.

Prevention

A large proportion of cases of HCV infection progress to chronicity suggesting an ineffective neutralizing host immune response, a conclusion also supported by data from experiments in chimpanzees; and preliminary results on the protection conferred by recombinant envelope proteins have been disappointing. An alternative approach using HCV cDNA vaccination, predicted to evoke both humoural and cell-mediated immune responses, is currently being adopted by a number of groups but no results are yet available.

A better understanding of the protective immune response in the cases where it can be unambiguously documented will be required before a potentially effective vaccine can be designed.

Treatment

The only treatment available is interferon-alpha which appears to control the infection in about 15% of the cases. In all the other cases, markers of infection are either unaffected by treatment or re-appear soon after cessation of treatment. Liver transplantation has been used with limited success since re-infection of the grafted liver is almost inevitable.

Following a better understanding of the molecular biology of HCV, it is expected that a new generation of antivirals directed at the NS3 protease and also at the RNA polymerase will soon become available. In the absence of a cell culture system to grow the virus, and in view of the restricted host range of HCV, the development of such antivirals is at present almost entirely dependent on *in vitro* systems based on recombinant HCV enzymes.

HCV and Hepatocarcinogenesis

An association between chronic HCV infection and hepatocellular carcinoma has been reported in several studies, in most cases in association with cirrhosis (IARC Monograph 1994). Whether the association results from the accumulation of mutations arising from the regenerative process or whether it is mediated by a specific virus-induced protein(s) is not known. Since the virus is an RNA virus and does not appear to have a DNA replicative intermediate as part of its replication strategy no direct mutagenic effect arising from integration of viral sequences is predicted.

Some Unanswered Questions

Further understanding of the HCV life cycle is greatly dependent on the availability of a cell culture system (or experimental animal model) that supports virus replication and active research is presently focussed on this aspect. Meanwhile the transfection of synthetic RNA of genomic polarity into suitable cells, if successful, is likely to generate important additional data both on the sequential processing of HCV polyprotein and on HCV replication. These studies will be complemented by improvements of current or development of new *in vitro* assays for NS3 protease, helicase, ATPase, and RNA polymerase activities. Availability of such HCV enzymatic assays will in turn stimulate the development of new antivirals, an area of acute clinical need.

Another important area is the clarification of the neutralizing host immune response, and of the individual roles of both humoural and cellular immunities. Such knowledge will become the basis for the rational design of a protective vaccine.

Immunological studies may benefit from the development of a virus-producing trangenic mouse and the use of VLP (Virus-Like Particles) or partially assembled complexes of native envelope proteins. These may prove to be effective antigens for the detection of putative, hitherto missed conformational epitopes and effective immunogens for inducing a protective response. Finally the results of DNA vaccination, as an alternative to conventional vaccination, is being awaited with great interest.

References

Abe, K., Kurata, T. and Shikata, T. *Arch. Virol.* 1989; **104**: 351–355.
Bartenschlager, R., Ahlborn-Laake, L., Mous, J. and Jacobsen. *J. Virol.* 1994; **68**: 5045–5055.

Bradley, D., McCaustland, K. and Krawczynski, K. *et al. J. Med. Virol.* 1991; **34**: 206–208.

Brown, E.A., Zhang, H., Ping, L-H. and Lemon, S.M. *Nucleic Acid Research* 1992; **20**: 5041–5045.

Chambers, T.J., McCourt, D.W. and Rice, C. *Virology* 1990; **177**: 159–174.

Chang, S., Yen, J-H. and Kang, H-Y. *Biochem. Biophys. Res. Comm.* 1994; **205**: 1284–1290.

Chen, P-J., Lin, M-H., Tai, K-F. *et al. Virology* 1992; **188**: 102–113.

Choo, Q-L., Kuo, G., Weiner, A. *et al. Science* 1989; **244**: 359–362.

Choo, Q-L., Richman, K.H., Han, J.H. *et al. PNAS* 1991; **88**: 2451–2455.

Colombo, M., Kuo, G., Choo, Q-L. *et al. Lancet* 1989; **ii**: 1006–1008.

Dubuisson, J., Hsu, H.H., Cheung, R.C. *et al. J. Virol.* 1994; **68**: 6147–6160.

Dubuisson, J. and Rice, C. *J. Virol.* 1996; **70**: 778–786.

Farci, P., Alter, H.J., Govindrajan, S. *et al. Science* 1992; **258**: 135–139.

Grakoui, A., Wychowski, C., Lin, C. *et al. J. Virol* 1993; **67**: 1385–1395.

Grakoui, A., McCourt, D.W., Wychowski, C. *et al. PNAS* 1993b; **90**: 10583–10587.

Harada, S., Watanabe, Y., Takeuchi, K. *et al. J. Virol.* 1991; **65**: 3015–3021.

He, L-F., Ailing, D., Popkin, T. *et al. J. Inf. Dis.* 1987; **156**: 636–640.

Hijikata, M., Kato, N., Ootsuyama, M. *et al. Biochem. Biophys. Res. Comm.* 1991; **175**: 220–228.

Hijikata, M., Kato, N., Ootsuyama, M. *et al. PNAS* 1991; **88**: 5547–5551.

Hijikata, M., Shimizu, Y.K., Kato, H. *et al. J. Virol.* 1993a; **67**: 1953–1958.

Hijikata, M., Mizushima, H., Akagi, T. *et al. J. Virol.* 1993b; **67**: 4665–4675.

Hijikata, M., Mizushima, H., Tanji, Y. *et al. PNAS* 1993; **90**: 10773–10777.

Hong, Z., Ferrari, E., Wright-Minogue, J. *et al. J. Virol.* 1996: **70**: 4261–4268.

Horzinek, M.C. in '*Non-arthropod-borne Togaviruses*' London Academic Press, 1981, pg40–41.

Kaito, M., Watanabe, S., Tsukiyama-Kohara, K. *et al J. Gen. Virol.* 1994; **75**: 1755–1760.

Kato, N., Hijikata, M., Ootsuyama, Y. *et al. PNAS* 1990; **87**: 9524–9528.

Kato, N., Ootsuyama, S., Ohkoshi, T. *et al. Biochem. Biophys. Res. Comm.* 1992; **189**: 119–127.

Kim, D.W., Gwack, Y., Han, J.H. and Choe, J. *BBRC* 1995; **376**: 221–224.

Kolykhalov, A., Agapov, E. and Rice, C. *J. Virol.* 1994; **68**: 7525–7533.

Kolykhalov, A.A., Agapov, E.V., Blight, K.J. *et al. Science* 277:570–574.

Komoda, Y., Hijikata, M., Sato, S. *et al. J. Virol.* 1994; **68**: 7351–7357.

Kumar, U., Cheng, D., Thomas, H.C. and Monjardino, J. *J. Gen. Virol.* 1992; **73**: 1521–1525.

Kumar, U., Monjardino, J. and Thomas, H.C. *Gastroenterology* 1994; **106**: 1072–1075.

Lanforfd, R., Chavez, D., Chisari, F. and Sureau, C. *J. Virol.* 1995; **69**: 8079–8083.

Lechmann, M., Ihlenfeldt, H.G. and Braunschweiger, I. *et al. Hepatology* 1996; **24**: 790–795.

Lin, C., Lindenbach, B.D., Pragai, B. *et al. J. Virol.* 1994; **68**: 5063–5073.

Lin, C., Pragai, B., Grakoui, J. *et al. J. Virol.* 1994; **68**: 8147–8157.

Lo, S.Y., Selby, M., Tong, M. and Ou, J.H. *Virology* 1994; **199**: 124–131.

Lohmann, V., Körner, F., Herian, V. and Bartenschlager, R. *J. Virol.* 1997; **71**: 8416–8428.

Miller, R.H. and Purcell, R.H. *PNAS* 1990; **87**: 2057–2061

Missale, G., Bertoni, R., Lamonaca, V. *et al. J. Clin. Invest.* 1996; **98**: 706–714.

Miyamoto, H., Okamoto, H., Sato, K. *et al. J. Gen. Virol.* 1992; **73**: 715–718.

Nasoff, M.S., Zebedee, S.L., Inchauspe, G. and Prince, A.M. *PNAS* 1991; **88**: 5462–5466.

Okamoto, H., Okada, S. and Sugyama, Y. *et al. J. Gen. Virol.* 1991; **72**: 2697–2704.

Pilipenko, E.V., Gmyl, A.P., Maslova, S.V. *et al. Cell* 1992; **68**: 119–131.

Prince, A.M., Huima-Byron, T., Parker, T.S. and Levine, D.M. *J. Vir. Hepatitis* 1996; **3**: 11–17.

Ralston, R., Thudium, K., Berger, K., *et al. J. Virol.* 1993; **65**: 6753–6761.

Ray, R.B., Lagging, L.M., Mayer, K. and Ray, R. *J. Virol.* 1996; **70**: 4438–4443.

Reed, K.E., Grakoui, A. and Rice, C.M. *J. Virol.* 1995; **69**: 4127–4136.

Rehermann, B., Chang, K-M., McHutchinson, J. *et al. J. Virol.* 1996; **70**: 7092–7102.

Saito, I., Miyamura, T., Ohbayashi, A. *et al. PNAS* 1990; **87**: 6547–6549.

Sakamuro, D., Furukawa, T. and Takegami, T. *J. Virol.* 1995; **69**: 3893–3896.

Sallberg, M., Ruden, U., Wahren, B. and Magnius, L. *J. Clin. Microbiol.* 1992; **30**: 1989–1994.

Santolini, E., Pacini, L., Fipaldini, C. *et al. J. Virol.* 1995; **69**: 7461–7471.

Selby, M., Glazer, E., Masiarz, F. and Houghton, M. *Virology* 1994; 114–122.

Shih, C.M., Miyamura, T., Chen, S.Y. and Lee, Y.H. *J. Virol.* 1993; 5823–5832.

Simmonds, P., Holmes, E.C., Cha, T.A. *et al. J. Gen. Virol.* **74**:661–668.

Steinkuhler, C., Urbani, A., Tomei, L. *et al. J. Virol.* 1996; **70**: 6694–6700.

Suzich, J., Tamura, J.K., Palmer-Hill, F. *et al. J. Virol.* 1993; **67**: 6152–6158.

Takahashi, K., Kishimoto, S., Yoshizawa, H. *et al.Virology* 1992; **191**: 431–434.

Takamizawa, A., Mori, C., Fuke, I. *et al. J. Virol.* 1991; **65**: 1105–1113.

Tanaka, T., Kato, N., Cho, M-J. and Shimotohno, K. *Biochem. Biophys. Res. Comm.* 1995; **215**: 744–749.

Tanji, Y., Hijikata, M., Hirowataru, Y. and Shimotohno, J. *J. Virol.* 1994; **68**: 8418–8422.

Tanji, Y. Kaneko, T., Satoh, S. and Shimotohno, K. *J. Virol* 1995; **69**: 3980–3986.

Tomei, L., Failla, E., Santolini, R. *et al. J. Virol.* 1993; **67**: 4017–4026.

Tsukyiama-Kohara, K., Iiauka, N., Kohara, M. and Nomoto, A. *J. Virol* 1992; **66**: 1476–1483.

van Doorn, L-J., Capriles, I., Maertens, G. *et al. J. Virol* 1995; 773–778.

Weiner, A.J., Brauer, M.J., Rosenblatt, J. *et al. Virology* 1991; **180**: 842–848.

Weiner, A.J., Geysen, H.M., Christopherson, J.E. *et al. PNAS* 1992; **89**: 3468–3472.

Yen, J.H., Chang, S.C., Hu, C.R. *et al. Virology* 1995; **208**: 723–732.

Yanagi, M., Purcell, R.H., Emerson, S.V., and Bukh, J. *PNAS* 1997; **94**:8738–8743.

Yuan, Z-H., Kumar, U., Thomas, H.C., *et al. Biochem. Biophys. Res. Comm.* 1997; **232**:231–235.

Yuasa, T., Ishikawa, G., Manabe, S-I. *et al. J. Gen. Virol.* 1991; **72**: 2021–2024.

PART 5

HEPATITIS E VIRUS

History

After HAV and HBV were identified the residual group of viral hepatitis became known as Non-A Non-B (NANB) hepatitis. One form of NANB hepatitis was not associated with blood or blood products, being more like hepatitis A, and appeared to be transmitted by the foecal-oral route (Balayan *et al.*,1983; Bradley *et al.*,1987). This type was reported to be associated with particularly severe forms of disease in pregnant women. Epidemics of this type of enterically-trasmitted Non A, Non B hepatitis were first reported in the Indian subcontinent but also in Southeast Asia and Africa. Sporadic cases were also described.

Although transmission and serial passage in cynomolgus macaques of the enterically-transmitted non-A, non-B agent was reported in association with the presence of 27–34 nm particles in faeces (Bradley *et al.*, 1987), the agent was ultimately identified through cloning of its genome in 1990 by Reyes *et al.* Using as the starting preparation a bile aspirate of high virus titre (1,000 particles/EM grid square) total RNA was extracted -as it was assumed to be a RNA virus- and cDNA synthesized using random primers. After cloning in bacteriophage lambda *gt10* using *EcoRI* linkers the library was differentially screened with radioactive total cDNA probes from both infected and non-infected bile. From sixteen clones originally isolated only one clone, ET 1.1, of about 1.3 kb, appeared to detect a unique band in the cDNA library. A probe made from this clone also hybridized specifically to cDNA prepared from RNA extracted from stools of putative epidemic non-A, non-B hepatitis patients from Somalia, Borneo and Pakistan, and to a 7.5 kb single-stranded RNA species extracted from infected liver. On the other hand no genomic sequences, human or from the cynomolgus macaque, were found to hybridize to the ET 1.1 cDNA. Finally the sequence from the ET 1.1 clone contained the sequence GDD, a conserved motif which is the hallmark for viral RNA polymerases.

As an alternative cloning strategy the bacteriophage lambda vector *gt11* was used for making an expression library and screening with convalescent serum from a Mexican patient led to the successful identification of clones specifically associated with epidemic non-A non-B hepatitis.

The agent was considered to be the virus responsible for the majority of enterically-transmitted (ET) Non A, Non B hepatitis and was designated Hepatitis E Virus (HEV) (reviews by Krawczynski, 1993; Bradley, 1995).

The Virus

Classification: The Caliciviridae

The virus has been included in the *Calicivirus* family based on morphological similarities with members of that family. However the genome organization shows a significant difference with caliciviruses in that the small ORF, ORF3, is located mostly within ORF2 but is found to be C-terminal in caliciviruses. Homologies with the alphavirus junctional sequences led to the suggestion that HEV may be a non-enveloped 'alpha-like' virus.

The Virion

The virion is a non-enveloped spherical particle of 30–32 nm diameter with spikes and indentations on its surface (Fig. 21), icosahedral symmetry of unknown detail, buoyant density of 1.39–1.40 g/cm^3 in CsCl and 1.29 g/cm^3 in potassium tartarate/glycerol and a sedimentation coefficient of 183S.

The Genome

Complete HEV genomic sequences have been obtained from different geographic locations, two from Burma (Reyes *et al.*, 1990; Tam *et al.*, 1991), one from Mexico (Huang *et al.*, 1992) and one from China (Aye *et al.*, 1992).

The genome is a 7.5 kb molecule of single-stranded RNA of plus polarity containing a 3′ polyadenylated tail (Tsarev *et al.*, 1992; Koonin *et al.*, 1992; Reyes *et al.*, 1993). Encoded proteins are contained in 3 partially overlapping

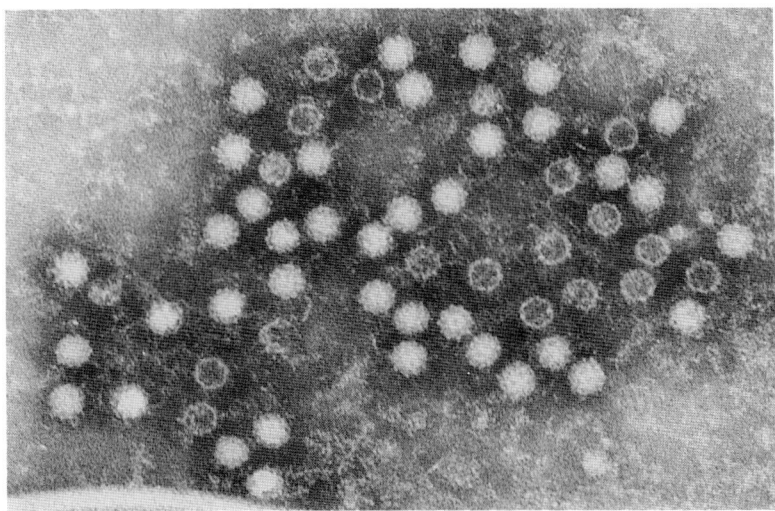

Fig. 21 Negative stain electron micrograph of Hepatitis E Virus (original image by E.H. Cook Jr., CDC; photograph courtesy of C.D. Humphrey, CDC) . . Magnification: × 181,440.

Open Reading Frames: the non-structural genes at the 5′ end and the structural genes at the 3′ end of the genome (Fig. 22).

The first ORF, ORF1, is 5,079 nucleotides long, starting 28 nucleotides from the 5′ terminus and terminating at nucleotide 5,107; ORF2 begins at nucleotide 5,147 and extends 1,980 nucleotides before terminating 65 nucleotides before the poly A tail; and ORF3 comprises 369 nucleotides, overlapping the last nucleotide of the ORF1 3′ terminus and substantially overlapping ORF2 by 328 nucleotides (Fig. 22).

HEV Proteins

Search for functional motifs in the 3 ORFS revealed that ORF 1 contains two such motifs characteristic of the nucleoside-triphosphate-binding activity associated with putative helicases (Koonin *et al.*, 1992). These motifs, which were conserved in other sequences, mapped to positions 975–982 and 1,029–1,032. Another motif, GDD, characteristic of viral RNA polymerases

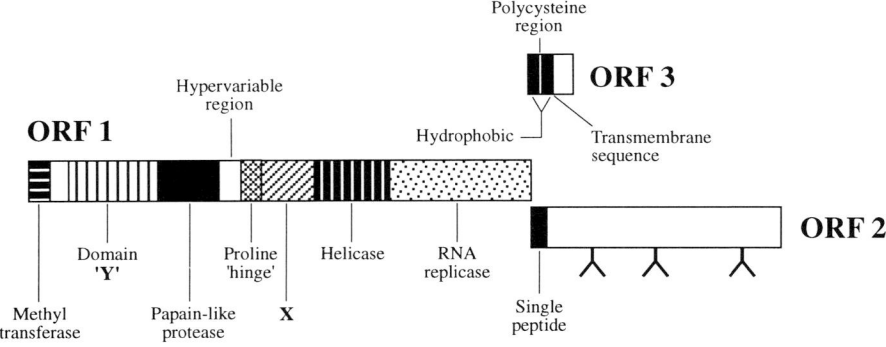

Fig. 22   HEV ORFs and functional domains. Y: glycosylation sites.

maps between positions 1,550 and 1,552. A putative methyltransferase, a "Y" domain of unknown function and a putative papain-like cysteine protease have also been described as has a polyproline region which is thought to act as a "hinge" between an upstream domain and the X region, a domain of unknown function.

The deduced gene order is 5′ methyltransferase, protease, X-gene, helicase, polymerase, capsid-3′.

The products of ORFs 2 and 3 have been recently characterized, after transfection of suitable recombinant plasmids into either COS or human hepatoma cells, or after *in vitro* translation of synthetic mRNA in the absence or presence of canine microsomal membranes (Jameel *et al.*, 1996). Pulse-chase experiments have shown the primary product of ORF 2 to be a precursor of about 82 kD which is subsequently processed to the mature product of 74 kD, in good agreement with the predicted size of the 660 amino acid ORF. Molecular forms with MWs of 82 kD and 88 kD which accumulate with longer pulses are thought to be glycosylated derivatives of pORF2 as indicated by their sensitivity to tunicamycin inhibition and endoglycosidase digestion. Three potential N-linked glycosylation sites (Asn-X-Thr/Ser) have been identified in ORF2 at positions 137, 310, and 561. The gp protein appears to form non-covalent homodimers but not to form heterodimers with the product of ORF3.

Due to its sensitivity to endoglycosidase H, gpORF2 is likely to contain high mannose residues, a modification which takes place in the endoplasmic reticulum and *cis* Golgi compartment. This observation implies a processing of gpORF2 through the ER and the *cis* Golgi towards the cell membrane. Furthermore the positively charged N-terminus of gpORF2 has been tentatively implicated in an interaction with the genome during the assembly of the particle.

The small ORF3 encodes a polypeptide of 123 amino acids. A protein of 13.5 kD, in good agreement with the predicted size, has been identified in the cytoplasm of cells transfected with the expressing recombinant plasmid. The protein is not processed or glycosylated. Two hydrophobic domains have been described in the N-terminus (Fig. 22). A host protein of MW31 kD has been found to be associated with the ORF3 product but its significance is unknown. In fact little information is available as to the role of the ORF3 protein and on whether it is a structural protein of the virion.

Transcription and Replication of HEV RNA

Two subgenomic polyadenylated mRNAs of 2.0 and 3.7 kb, possibly coterminal with the 3′ end of the HEV genomic transcript have been identified on Northern blots from a Mexican HEV isolate (Tam *et al.*, 1991) and from a poly(A)-selected RNA from HEV-infected cynomolgus monkey liver (Yarbough *et al.*, 1991). Although the two transcripts were involved in translation of ORFs 2 and 3, it has not been established which transcript was involved in the translation of which ORF. The mechanism for generation of these subgenomic transcripts is not clear but it may be similar to the one proposed for Sindbis virus (alphavirus) where transcription is regulated by newly synthesized and processed viral proteins. Replication on the other hand is thought to involve the synthesis of minus strand and the formation of a double stranded replicative intermediate.

After virus entry and uncoating replication starts with genome translation of the non-structural genes (which include the RNA polymerase) followed by the assembly of a transcription complex involving the non-structural proteins in *cis* (and possibly host factors), which recognizes a promoter-like sequence for the synthesis of the minus strand. This in turn will become the template for the synthesis of progeny plus strands and of the subgenomic mRNAS for structural viral proteins. Recognition of other promoter-like sequences for the transcription of these 'late' mRNAs is thought to arise from further processing of the non-structural proteins now acting in *trans* in a newly assembled complex.

Pathogenesis

Histological changes characteristic of cholestatic hepatitis, i.e., obstruction of the bile canaliculi leading to bile stasis and of classical hepatitis with its necro-inflammatory changes have been described in HEV infection (Krawczynski, 1993). The latter includes an inflammatory infiltrate of mononuclear cells, macrophages and lymphocytes, with focal intralobular necrosis and acidophilic degeneration of hepatocytes. HEV Ag has been detected in the cytoplasm of infected hepatocytes of non-human primates 10 days after intravenous inoculation and remains detectable for 2–3 weeks. Some patients showed severe acute hepatitis with extensive hepatocyte necrosis which was fatal in some cases.

Animal Models and *In Vitro* Culture of HEV

Cynomolgus macaques have been shown to be highly susceptible to human HEV isolates but other primates including owl, rhesus and African green monkeys as well as chimpanzees born in captivity are also susceptible, although chimpanzees less than the others. Virus has been produced after infecting these animals, bile being a particularly good source of virus, and antisera reactive with HEV particles have been obtained.

Propagation of HEV in a variety of cell lines (BHK-21, Vero, LLC-MK$_2$, BSC-1, FRhK-4, HEK and AGMK) has proved unsuccessful. Claims that 2BS and A549 cells support virus growth await confirmation from re-infection of fresh cultures with putatively infected culture medium and detection of the replicative intermediate antigenomic (minus) strand. A recent study describes the propagation of HEV-infected cynomolgus macaque's hepatocytes in long-term cultures and documents the presence of viral single-stranded RNAs of both polarities in the absence of CPE. The availability of cell culture systems for propagation of HEV would constitute a significant contribution to the study of the molecular biology of this virus and the testing of potential antiviral compounds.

The Disease

Natural History of the Disease

The disease is characterized by an acute infection which is resolved and appears to leave lasting immunity (Krawczynski, 1993). In rare cases the acute infection is associated with massive liver necrosis that results in death as described in young women where case-fatality rates can reach 15%–25%. The virus is thought to be water-borne and to enter the bloodstream through the gastro-intestinal tract. The viraemia is shortlived and the virus starts being eliminated in the bile (and faeces) before symptoms develop or serum liver enzyme levels (ALT, AST) become elevated. The incubation period is approximately 5–6 weeks. HEV RNA can then be detected briefly in serum and also found in stools. HEV antigen is detected in the cytoplasm of the hepatocyte by immunofluorescence (as seen in the cynomolgus macaques) and virus particles in the stools by immunoelectron microscopy. Antibodies become detectable, IgM first around the time of presentation and IgG about two weeks later; neutralization of the virus with resolution of the infection occurs about 5 weeks after the start of symptoms.

Different recombinant antigens and synthetic peptides have been used as the solid phase of Western blot-based, EIA and other HEV antibody assays. The antigens used are recombinant fusion proteins that include either the carboxyl two thirds of ORF2, complete ORF2 or portions of ORF3 and ORF2. Since cross-challenge studies have demonstrated the existence of a single serotype of HEV recombinant protein derived from virus from any geographical location can be universally used for diagnosis and for the development of a vaccine.

HEV Ag assays include the immunostaining of biopsy speciments with convalescent serum and immunoelectron microscopy in faecal specimens.

Finally HEV RNA can be detected both in serum and in faeces by RT-PCR using HEV specific primers from conserved regions of the genome.

Epidemiology

Outbreaks of Hepatitis E have been reported in many different countries of Asia, Africa and Central America. Recurrent epidemics appear to have a periodicity of 5–10 years and the reservoir of the virus between epidemics is unknown. Many of the outbreaks tend to follow heavy rains.

The virus seems to infect particularly persons of 15–40 years of age and although a fecal-oral route has been established, person-to-person transmission is uncommon, ranging from 0.7%–2.2% even amongst communities with very poor sanitation facilities. This compares with 50%–75% in households with hepatitis A.

Seroprevalence studies in countries where HEV is endemic shows that anti-HEV positivity is less than 5% in children under 5 years of age and increases to 10%–40% among those over 25. This is again in marked contrast with Hepatitis A which shows a seroprevalence of more than 90% in children under 5 years old in countries where the infection is endemic.

Seroprevalence rates of 1–2% have been reported in industrialized countries amongst individuals who have not travelled abroad but the significance of such findings is open to question in view of the significant variation in sensitivity and marked discordance between the various available assays.

Prevention

Improvements in socio-econonomic conditions, particularly in sanitation, and also in health education, are likely to be the most significant factors in the control and ultimate eradication of this infection in the countries where it is most prevalent.

A preliminary report suggesting that a fusion protein (*trp*-HEV), offered protection against virus challenge of cynomolgus macaques has not been confirmed nor is there information on subsequent vaccine development.

In view of the low morbidity and mortality of this disease there appears to be little justification for the development of a vaccine for mass vaccination in countries with high HEV prevalence. Its potential additional use for the protection of travellers to those areas seems equally unjustified on the basis of cost effectiveness.

In groups with high mortality, i.e., pregnant women, acquisition of passive immunity (by administration of immune serum or neutralizing anti-HEV immunoglobulins) is expected to prevent the infection or else to markedly attenuate its clinical course.

Treatment

No specific treatment is available and apart from the serious clinical course in pregnant women the infection is rarely fatal and invariably resolves within 1–2 months after onset. The cause for the severe morbidity in pregnant women is unknown.

Some Unanswered Questions

The next few years will no doubt witness significant advances in our understanding of this virus, both at the cellular and molecular level.

The chemical composition and architecture of the virion are still unsolved and will require the purification of significant amounts of virus, either from the bile of infected cynomolgus monkeys or from infected cultured cells, if a system is found in which the virus can be efficiently propagated.

Much is to be learnt about the molecular biology of HEV since little is known about the transcription or the translation of the viral genome. The translation of the infecting genome, with the processing of the non-structural polyprotein and its regulation, and the transcription of the minus strand and of subgenomic RNAs are all challenging questions; as indeed are the mechanism of virus entry, the process of assembly and the release of the newly-synthesized virions.

In the areas of pathogenicity of HEV and management of HEV hepatitis, further elucidation of the underlying pathogenetic mechanism and improved diagnostic assays are important subjects for future research. Both the uderstanding and prevention of mortality in pregnant women are questions of significant scientific interest and major practical importance.

Finally it is predicted that a significant reduction in HEV seroprevalence will follow much needed socio-economic development in the countries currently worst affected by HEV epidemics. In this context the development of a vaccine for use in these countries does not appear to be a priority.

References

Aye, T.T., Uchida, T., Ilida, F. *et al. Nucleic Acids Res.* 1992; **20**: 3512.

Balayan, M.S., Andjaparidze, A.G., Saviskaya, S.S. *et al. Intervirology* 1983; **20**: 23–31.

Bradley, D.W., Krawczynski, K., Cook, E.H. *et al. PNAS* 1987; **84**: 835–845.

Bradley, D.W. *J. Hepatol.* 1995; **22**(suppl.1): 140–145.

Bi, S.J., Purdy, M.A., McCaustland, H.S. *et al. Virus Research* 1993; **28**: 233–247.

Huang, C.C., Nguyen, D., Fernandez, J. *et al. Virology* 1992; **191**: 550–558.

Jameel, S., Zafrullah, M., Ozdener, M.H. and Panda, S.K. *J. Virol.* 1996; **70**: 207–216.

Khuroo, M.S. *Am. J. Med.* 1980; **68**: 818–823.

Khuroo, M.S., Teli, M.R., Skidmore, S. *et al. Am. J. Med.* 1981; **70**: 252–255.

Krawczynski, K. *Hepatology* 1993; **17**: 932–940.

Panda, S.K., Nanda, S.K., Zafrullah, M. *et al. Clin. Microbiol.* 1995; **33**: 2653–2659.

Reyes, G.R., Purdy, M.A. and Kim, J.P. *Science* 1990; **247**: 1335–1339.

Reyes, G.R., Huang, C.C., Tam, A.W. and Purdy, M.A. *Arch. Virol.* 1993; **7**: 15–25.

Tam, A.W., Smith, M.M., Guerra, M.E. *et al. Virology* 1991; **185**: 120–131.

Tsarev, S.A., Emerson, S.U., Reyes, G.R. *et al. PNAS* 1992; **89**: 559–563.

Yarbough, P.O., Tam, A.W., Fry, K.E. *et al. J. Virol.* 1991; **65**: 5790–5797.

PART 6

OTHER HEPATITIS-ASSOCIATED VIRUSES: HGV/GBC

History

A group of hepatitis cases, potentially viral, remains after hepatitis viral hepatitis agents A-E have been excluded. Attempts to isolate a new hepatitis agent(s) responsible for such cases have resulted in the identification of a new human agent, HGV or GBC, as well as of two others, GBA and GBB, which appear to infect only marmoset monkeys (reviewed by Karayiannis and McGarvey, 1995). The significance of the human agent as a human pathogen remains to be established.

GB agents are so designated because they are thought to originate from an original inoculum derived from a surgeon (initials GB) infected with sporadic non-A non-B hepatitis in 1967. Transmission to marmoset monkeys of the species *Sanguinus* (tamarins) was successfully achieved by intravenous injection and raised serum transaminases were documented after 16–40 days and on five subsequent serial passages. Histological changes observed in the livers consisted of significant necro-inflammatory changes compatible with a diagnosis of viral hepatitis (Deinhardt *et al.*, 1967).

Following these observations further studies raised doubts as to the human origin of these agents and suggested instead that the changes observed in the marmosets had arisen from the reactivation of endogenous agent(s) (Parks and Melnick, 1969).

A renewed interest in this agent (s) ocurred when in 1993 cloning from a tamarin, acutely infected with the GB inoculum, led to the identification of two agents GBA, GBB (Simons *et al.*, 1995). Since none of these agents were found in human hepatitis cases by RT-PCR of serum RNA it was concluded that they are not human pathogens. However, by using degenerate primers from one region of the genome, a new agent, GBC, was subsequently found in humans (Simons *et al.*, 1995). Independent of these studies and utilizing a conventional cloning strategy from sera derived from non-A, non-B, non-C cases of human hepatitis, another group of workers at Genelabs also reported the identification of a new agent, which they designated HGV (Linnen *et al.*, 1996). Sequence data has since shown both agents, HGV and GBC, to be the same.

The Virus

Classification

The virus, and the tamarin viruses, are flaviviruses which appear to form a closely related cluster that includes hepatitis C virus (Fig. 23).

The Virion

The virus particle has not been characterized by electron microscopy or by ultracentrifugation.

The Genome

Two full genomes were originally reported by the group at Genelabs (Linnen *et al.*, 1996). Both were derived from human sera: the first from a patient with hepatitis originally thought to be unrelated to hepatitis viruses A-E (but later

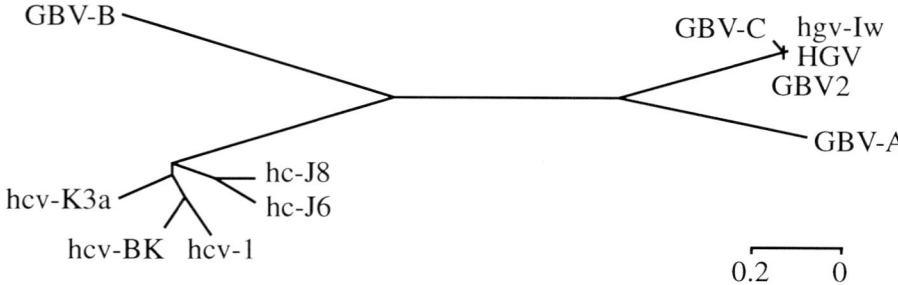

Fig. 23   Dendogram showing genetic relatedness between several recently identified marmoset and human-derived flaviviruses.

found to be positive for HCV by a second generation assay) and the second from an asymptomatic individual with a history of intermittently raised transaminases. Cloning in the original case was performed in bacteriophage *gt11* and screening carried out with serum from the same patient. In addition to HCV clones other sequences were obtained which were unrelated to any previously known published sequences or to genomic sequences from *E. Coli*, *S. cereviseae* or human origin. Furthermore the cloned sequences (clone 470-20-1) were present in the cloning source serum with a titre of about 10^6 genomes/ml but absent in healthy control sera. Extension of the original sequence was achieved from the original cDNA by means of overlapping sequences generated using anchored PCR amplification.

The full sequence was 9392 nucleotides long and includes an ORF spanning almost the full length of the genome and encoding a polyprotein of 2873 amino acids. There are two untranslated regions flanking the ORF, 5' and 3' UTRs of 458 and 315 nucleotides respectively. The HGV polyprotein showed 43. 8% homology with the tamarin derived GBA and 28.4% and 26.8% homologies with GBB and HCV respectively. Homology with GBC sequence (see below) was 85.5% at nucleotide level and 100% at amino acid level indicating that they are isolates of the same agent.

The organization of the genome is similar to that of HCV with the structural genes clustered at the aminoterminus of the polyprotein and the non-structural genes towards the carboxyl end. Amongst the latter motifs for a helicase and serine protease (NS3) and a RNA-dependent RNA Polymerse have been identified. The carboxy terminus of the E1 gene and amino-terminus of E2 have been clearly mapped and this has helped locate the NH_2-terminus of the HGV polyprotein relative to that of HCV. This shows the aminoterminal region of HGV to be variable between isolates and much shorter than that of HCV (Linnen *et al.*, 1996) and suggests that the core protein is either truncated or absent. Surprisingly, the original immunodominant epitope contained in clone 470-20-1 is encoded in a short 119 amino acid ORF in the complementary strand.

The second genomic sequence reported by the same group was found to be 9103 nucleotides in length and to encode a longer polyprotein of 2910 amino acids (Linnen *et al.*, 1996). The two sequences show an homology of 90.5% at nucleotide and 97.5% at amino acid level.

A different near full-length sequence, that of GBV-C, has also been reported (Leary *et al.*, 1996). This was achieved after extending the original sequence obtained from the serum of a patient from West Africa by RT-PCR amplification of RNA using degenerate primers derived from the helicase regions of GBA, GBB and HCV. The sequence was extended to 9,125 nucleotides. Starting with the first methionine, the long ORF consists of 2874 amino acids although the initiator methionine still has to be confirmed. The length of the putativa 5' untranslated region (5' UTR) is 439 nucleotides and the 3' UTR is 61 nucleotides long and does not include a homopolymeric region. The ORF of this sequence was compared at the amino acid level with the ORFs of GBV-A and GBV-B and HCV1 and found to be more related to GBV-A (48%) than to GBV-B and HCV (approx. 30%) and no more related to HCV than GBV-A or GBV-B (Leary *et al.*, 1996). Significant if limited aminoacid homologies between GBV-C and HCV in the putative NS3 and NS5B regions suggest that these are similarly responsible for encoding a serine protease and the RNA polymerase respectively. The serine protease triad motif of His, Asp and Ser has, in fact, been mapped to positions 1013, 1037 and 1094 as has the polymerase signature sequence GDD to amino acids 2649–2651 (Leary *et al.*, 1996). Although studies on polyprotein processing have not been published to date, putative cleavage sites for both host and virus-coded proteases have been identified. Four signalase sites have been mapped in the putative structural region in comparable positions to those seen in HCV. Within the envelope region three putative host signalase sites would generate E1, E2 and p7 similarly to what is seen in HCV. As to the processing at the NS2/NS3 boundary a second viral protease appears to be involved and subsequent processing downstream, i.e., NS3/NS4A, NS4A/NS4B, NS4B/NS5A and NS5A/NS5B is presently being postulated to be carried out by the NS3 protease in a similar manner to that documented for HCV.

A fourth sequence has recently been reported from Japan (Shao *et al.*, 1996). The patient was HGV RNA positive by RT-PCR and negative for markers of HAV, HBV and HCV. The sequence is 9375 nucleotides long and the ORF encodes a polyprotein of 2873 amino acids. A significant number of nucleotide changes between the Japanese isolate and GBV-C were found throughout the genome, including the 5' UTR . These ranged from 13.8% (86.2% homology) when compared with GBV-C, to 6.7% (93.3% homology) with HGV. This additional sequence provides further confirmation of the non-existence of a hypervariable region in HGV in contrast to HCV. In spite of the diversity homologies at amino acid level are very high (96.1%–100%) between the Japanese and the three previously reported sequences. Phylogenetic analysis at the amino acid level of the Japanese isolate (hgV-Iw) shows that it is more closely related to the HGV isolates from the USA than to those from West Africa. Such analysis between the four HGV-C sequences and GBA, GBB and several HCV isolates is shown in Fig. 23.

Viral Proteins

Although most proteins have been expressed in a variety of vectors, the sequential processing of the polyprotein remains to be elucidated.

Replication and Pathogenesis

Little is known about the replication of the virus which has not been shown to infect any cultured cells. Markers of liver replication have also not been reported.

The pathogenic role of the virus has been questioned since it has not been demonstrated that it is causally associated with liver cell injury in human hepatitis and has failed to cause clinical hepatitis in chimpanzees. No antibody responses to a variety of recombinant viral proteins have so far been documented in the serum of patients who are RT-PCR positive for HGV/GBC RNA. Surprisingly, this also includes antibodies to the immunodominant epitope encoded in the antigenomic strand of the clone 470-20-1 which led to its identification.

Epidemiology

Although HGV/GBC RNA has been found in sera from all over the world, the number of cases of hepatitis where it is present in the absence of other hepatitis viruses, amounts to less than 10% of this residual group. Even in this small group its role as the pathogen implicated remains to be established. HGV RNA has been detected in a significant percentage of between 1% and 2% of voluntary blood donors in the United States, a prevalence which is significantly higher than that of HCV (Alter, 1996). Patients with parenteral exposure to blood, i.e., patients on haemodialysis, haemophiliacs and intravenous drug abusers show high prevalence of HGV. In one prospective study of the recipients of HGV RNA positive blood, 75% were shown to have no biochemical evidence of liver disease and in the few cases where HGV was the only agent identified, transaminase levels were very modestly raised and bore no relation to HGV RNA (Alter, 1996). In another study from Japan (Masuko *et al.*, 1996) where HGV RNA prevalence in patients on dialysis is high (3.5%), no raised transaminase levels were detected in association with HGBV-C (HGV) persistence over 16 years. In fact the discrepancy between the significant HGV RNA positivity in the general population and the very small number of cases where it could be causally related to demonstrable clinical hepatitis suggests either very low or no pathogenicity. The agent can be almost seen as an innocuous marker of a previous transfusion or parenteric contact with blood or blood products.

Relationship with GBA

Although GBC was cloned from an African patient who was part of a group reportedly antibody-positive for GBA no confirmatory evidence of GBA as a human pathogen has been published since. Likewise no GBC infection of tamarins has been reported. Thus, the accumulated evidence to date strongly indicates that GBA and GBB are tamarin viruses and that GBC is a human virus. False positivity of the anti-GBA assays and partial homology between the genomic sequences in the regions of the PCR primers are possible explanations for the findings originally reported.

References

Alter, H. *N. Engl. J. Med.* 1996; **334**: 1536–1537.
Deinhardt, F., Holmes, A.W., Capps, R.B. and Popper, H.J. *Exp. Med.* 1967; **125**: 673–681.
Karayiannis, P. and McGarvey, M.J. *J. Viral. Hepatitis* 1995; **2**: 221–226.
Leary, T.P., Muerhoff, A.S., Simons, J.N. *et al. J. Med Virol.* 1996; **48**: 60–67.
Linnen, J., Wages Jr., J., Zhang-Keck, Z-Y. et al. *Science* 1996; **271**: 505–508.

Masuko, K., Mitsui, T., IWano, K. et al. *N. Engl. J. Med.* 1996; **334**: 1485–1490.

Parks, W.P. and Melnick, J.L. *J. Infect. Dis.* 1969; **120**: 539–547.

Shao, L., Shinzawa, H., Ishikawa, K. *et al. Biochem. Biophys. Res. Comm.* 1996; **228**: 785–791.

Simons, J.N., Pilot-Mathias, T.J., Leary, T.P. *et al. PNAS* 1995; **92**: 3401–3405.

Simons, J.N., Leary, T.P., Dawson, G.J. *et al. Nature Medicine* 1995; **1**: 564–569.